WOMEN, AGEING AND ARCHETYPES:

Ideas, Images and Support on the Post-Menopausal Journey

by

Margaret Stone

Written by Margaret Stone

ISBN 978-1-904098-48-5

Published by:
Wildtrack Publishing,
Venture House,
103 Arundel Street,
Sheffield S1 2NT

Typeset and processed by Christine Handley

CONTENTS

Preface

Ageing can be a difficult time for anybody. It can mean facing changes that we don't want, some of them frightening. Ageing is especially difficult for women. There are few role models of powerful older women, and we still feel pressure on us to conform to unreasonable standards of feminine beauty or maternal usefulness. Getting old can make us feel useless and redundant, unattractive and even a figure of scorn. We may also fear worse times ahead. Small wonder then that a multitude of books exist telling us how to feel, look and act younger than we are.

Conscious ageing is a new movement which encourages us to look ageing full in the face. We are to be fully conscious of what is going on, rather than denying it. By doing so we can not only accept what we are but work towards the rich rewards that can go with a good third age: wisdom, maturity, meaning, purpose, spirituality, serenity, humour and a measured self esteem.

For each individual, ageing is an exploration into new territory. It can often be a dark experience, as our bodies change and we face our new limitations, including the knowledge of our mortality. Such a journey requires each of us to have the courage to be the Hero of our own journey. The Hero is an archetypal figure: in other words, one of enduring significance that is found in all cultures. Everyone knows what a Hero is, although s/he is usually seen as male. We need to reclaim the Hero as part of our own psyche.

As we go through life our identity is constantly changing. Sometimes the changes are abrupt and difficult to negotiate, bringing the danger of losing our sense of meaning and purpose. Telling the story of our life as that of the Hero on her archetypal journey, rather than as helpless Victim of random fate, will influence how well we live out our third age.

When we go far back into prehistory to the origins of humanity, and understand the reasons for the evolution of the menopause and of women's particular pattern of ageing, we see that our loss of oestrogen is not just some "deficiency disease". Instead the menopause was a positive step forward in evolutionary terms, and was directly connected to our role as grandmother of the tribe, source of wisdom and nourishment – the Wise Woman, the second major archetype to be explored.

Goddess worship has been widespread throughout human history, yet the images, roles and status of woman have been distorted and suppressed. There are serious questions to explore before we can embrace this third major archetype. Why is it that such powerful archetypes have been so nearly lost to women? Are they truly so powerful, or are ideas of women Elders and the Great Goddess just wishful thinking?

By looking in detail at some of the evidence of what was going on in our history and prehistory, we can be more confident in reclaiming the Goddess as our charter myth. We can distinguish between literal truth and metaphor. Then we can look to the varieties of goddesses in the various myths and representations to help us reclaim our own sense of power and strength.

Courage and the search for meaning (Hero), wisdom (Wise Woman) and power (Goddess) will give us the strength to look into what the future holds for us, including journey's end - death. But the Hero on her journey will have times when help is needed. As well as exploring the use of various archetypes to help us through the dark times, we explore what tools might help us on our journey and how we can find emergency help and comfort when most in need.

We may live out more than a third of our life after the menopause. There is no reason why those years should not be rich, rewarding and enjoyable, and an example for those women following on behind us.

CHAPTER 1 – Ageing and How We Define Ourselves

Women's lives have traditionally been divided into the ages of Maiden, Mother and Crone, which can be related to the waxing, full and waning moon. These divisions reflect not only the reproductive stages of our lives, but also the main stages of our own journey of personal identity. The more we are happy with who we are, and the more sympathetic our environment, the smoother our journey will be. Even then it may not be easy.

The time of the Maiden – the waxing moon - is a time of freedom for developing our potential and trying out different pathways before we decide on our way forward. We may be held back in this stage both by society and by our own self doubts so that we do not blossom as we should. Motherhood – the full moon – is the peak time of love and work (as Freud summed up the tasks of life) and may or may not involve actual childbearing. It is a time which today involves much effort and responsibility and can feel exhausting.

Then the menopause – the time of the shutting down of our reproductive capability – heralds the stage of the Crone, the waning moon. This is when we should reap the harvest of our earlier years. Ideally this is the time of growing wisdom and maturity, perhaps the delights of grandmotherhood, a time to turn to new interests and devote more time to our personal and spiritual needs. But it can be burdened with health and money worries and a low status in society.

Since women's bodies are so much more tied up with reproductive stages than men's, we tend to be much more conscious of the passing of the years. This can give us an advantage in terms of personal growth, but it also has its downside. Ageing ain't for cissies, as the old saying has it.

The menopause itself can cause a great deal of mental and physical distress as our body struggles to make huge adjustments in a short space of time. At the same time our family responsibilities may actually increase instead of decreasing, with difficult demanding teenagers and increasingly frail parents. Or we may be grieving as we become middle-aged orphans. Modern patterns of employment don't allow for any slacking off and we may be working harder than ever.

We may have worries about our health as the effects of ageing start to become noticeable. It can be upsetting as we see ourselves becoming less sexually attractive, and we all have to mourn the loss of our power to give birth, even if we hadn't planned to do so. We may become truly aware of our mortality, and that of our loved ones, for the first time. As we realize that we too will grow old, we have to mourn lost opportunities, regret mistakes made that cannot be undone, and accept that certain ambitions will now never be realized. We may worry about the future, about inadequate pensions, and about who will care for us. We may fear the approaching end of life.

However, consciousness of ageing can also make us more determined to make the most of our lives. The ways of living that used to content us are no longer good enough. We don't feel like putting up with inconsiderate husbands, unappreciative bosses, ungrateful relatives, unsupportive friends. We become more selfish, in the sense of wanting to attend to our own needs rather than always put others first. Projects and ideas that we have left undeveloped for years may now become a priority.

We may find that all of a sudden, what used to enthuse us no longer does so. The particular ambitions we used to have are now gone. The job we trained for no longer interests us. The kind of people we used to feel comfortable with has changed. We're not sure what we want to do, or even who we are. We can feel like we are floundering.

We all have to go through the physical and mental adjustments of ageing, and because women are strong, pretty well all of us do so successfully, one way or another. Any ways in which we can increase our inner reserves of strength, hope, optimism and determination will help us to come through with more laughter and less pain. One major factor that will affect this is our own set of beliefs and images about older women.

This is a time when we have to redefine ourselves, even if only in terms of our physical being, and it is important that we choose the right positive images with which to do so. We have to choose our images of who we are, and not let negative images be imposed upon us.

Archetypes are universal images or roles that are found in every society, in similar form, because they address a universal human condition. Archetypes are brain-based and arose through our physical evolution. They may be suppressed in particular times and places but can be called forth in the right conditions. Some examples are the Leader, the Warrior, the Wise Woman, the Hero, the Joker, the Shaman. Women encompass Maiden, Mother and Crone as they age but beyond that different women may embody different archetypes according to their nature and their destiny.

As we meet the difficulties of ageing we may respond as one of two archetypes: Victim or Hero. The Victim will suffer passively or waste her energy in bewailing her fate. The Hero will know that when embarking on an adventure or a Quest, there will be times full of dread, but a Hero will win through, slay the dragon and win the boon in the end, to the advantage of all around her.

As we age, we are on an epic journey that has been shaped by millions of years of human evolution. Our own personal journey, of which we are the Hero, reflects that of all women through the ages. The strength and love of these women was acknowledged in the archetypes found in myth and worship, and our sisters of the past can continue to be an inspiration to us. In this book we will explore many stories and female archetypes – heroes, wise women and goddesses – to help us on our way.

Whether or not we enter the third age with positive images of older women will partly depend on luck. We are shaped by the home in which we grow up.

You may want to take some time to think back to your childhood and the older women that you knew. How did they look and act, how was their health and financial status? How were they treated by other people? Were they respected? Did they seem at ease with themselves, were they happy?

Your mother's story in particular – if she lived long enough – will be important to you, since we can often expect to be like her, or at the other extreme we may be determined to be entirely different. Has she provided you with a positive or negative model of ageing? If the model is negative, can you see how your story might be different?

The various influences which shape our ideas and expectations also come from people we have never met: advertisers and television producers, writers and film directors, politicians and magazine editors. Even the glance of strangers in the street can carry judgement, helping you to feel a somebody or a nobody. As social beings, we are to a great extent defined by our relationship with others and by society at large.

As women we are seen as generally less important than men; as older women, we can fade almost into invisibility. We do have some positive role models today, but on the whole the current public image of older women will not help us to feel strong and powerful. In our quest to redefine ourselves at a transitional point in our lives, we need to be alert to how others seek to define us. If we are aware of the negativity around us, we can consciously work to stop it seeping into our soul, or to uproot it where it has already taken hold.

Age is not seen so negatively in men. Almost all the holders of power and wealth today are male. Generally speaking, the CEO of a multinational organisation, a Government leader, the Director General of the BBC or a newspaper magnate are all likely to be male. Not only that, but they are likely to have reached middle age, and may even be at retirement age. They may well have wrinkled skin and white hair.

These men are not objects of derision and pity – power and wealth completely change the stereotypes of ageing. Wrinkles and white hair now become the badges of wisdom, experience and maturity. Think of Merlin in the legends of King Arthur, or Gandalf in *Lord of the Rings*. Our most prominent figures in the media – newsreaders, weather men, documentary makers and presenters – are often middle aged when male, whereas when a woman is given such a role she is usually young and beautiful.

Men who do not have any power and wealth may well find old age difficult. Even in the prosperous west, many struggle on an inadequate pension and fear for their future. And the wealthy and powerful men will have their own fears about loss of virility, younger men snapping at their heels, and the ultimate mystery of death. They may even spend a great deal of money trying to stay young. However men are reassured that it is natural, it is OK, it is possibly even rewarding to be old, in a way that women are not.

Simone de Beauvoir in *The Second Sex* introduced the idea that we are taught – absorb from our culture - an idea about what being a woman means, which will be different in different places and times. We live in a society where on the whole, women are of secondary importance to men, have less power than men, and less say in what happens. This situation is often referred to as patriarchy (literally "father rule").

De Beauvoir suggested that in our culture there is a Self or Subject who is active and knowing, and who is assumed to be male. The common viewpoint – the view of "the man in the street" – is assumed to be a male view. The person reading the newspaper or watching a film is assumed to be male; an anonymous person is usually referred to as "he" rather than "she". The whole species is referred to as Man.

Women are then seen as the Other – different from men, and a projection of everything that the Self/Subject would hate to be and rejects for himself. Thus she is passive, voiceless. As a girl grows up, she learns that she must renounce trying to be a Subject (which will be difficult anyway as society will constrain her) in order to be desired as an Object or Other. What is important is not how she sees other people, but how they see her.

Women are the Object of the male Subject's gaze; women are taught see themselves that way, rather than seeing men as the Object of a female gaze. In real life or in film, a "male gaze" is assumed. Women may want to be seen as conforming or they may want to be seen as feisty, but they experience themselves *as being seen* and generally they want to look good and gain approval. Not being seen would really be failure.

Take this passage from *Northanger Abbey*:

> "She *was* looked at, however, and with some admiration; for in her own hearing two gentlemen pronounced her to be a pretty girl. Such words had their due effect; she immediately thought the evening pleasanter than she had found it before; she felt more obliged to the two young men for this simple praise than a true quality heroine would have been for fifteen sonnets in celebration of her charms, and went to her chair in good humour with everybody, and perfectly satisfied with her share of public attention." (Austen, 1818 p24)

Things haven't changed much in the two hundred years since Jane Austen wrote these words about a teenage girl.

Simone de Beauvoir suggested that one is not born but becomes a woman – that we learn the role of being a woman, with all its rules and restrictions as the Other. We may feel that we are much freer than we used to be – 'girl power' for example – but in fact we are hedged about all the time, as indeed are men, who also have to learn to be men. Women hate sitting by the phone, hoping that *he* will ring. He may be wishing it wasn't his job to find the courage to pick up the phone and risk rejection.

When roles are reversed in comedies the idea of male passivity and the pursuing female seems amusing, because female passivity seems natural. When women are too independent they may be accused of being "ball breakers". Women's status in the West has improved over the years, as women can achieve financial independence and all that goes with it, but the basic experience of being the object of men's attention lingers on.

In *The Coming of Age* de Beauvoir developed the idea of the Other to explore the question of women and ageing. She said that we have another Other within us – the older woman who is not our Self, but is the woman that others see. The older woman knows that others perceive her body as declining after the menopause, and she experiences this as a loss of femininity. According to de Beauvoir we capitulate to becoming the image which others provide, so if the Subject's gaze now sees us as undesirable – or now fails to see us at all – we become nobodies, unimportant, perhaps living through our children.

This is a very depressing view, and Germaine Greer in *The Change*, her book about the menopause, suggests that de Beauvoir's horror of old age was due to the miseries that can grip some of us during the menopause, the feelings of anxiety, depression, even doom at times. Also, again, on the whole things are supposed to be better today for women in the West than they were in the 1950s and 60s.

However, it is interesting to read Heather Dillaway's article on *Changing Menopausal Bodies* (2005). She interviewed 61 women in the United States – many middle class, some working class, mainly white but some black, about what the menopause has meant to them.

The modern women in Dillaway's survey see a great deal of continuity as they go through the menopause. Many are still working, some have delayed childbearing and so are still busy mothers. Some said they did not see the menopause as particularly significant. However two-thirds saw the physical changes as problematic and negative, and what they disliked most was the change in their appearance.

Dillaway talks of how the required look in the US (for the male gaze) is thin, young, white, well off, with no noticeable imperfections or disabilities. Women are required *not* to change if they are to remain acceptable. Dillaway sees this in de Beauvoir's terms as the menopausal woman becoming "different" from a culturally defined woman – reproductive, fertile, sexual, attractive - and crossing over into an "other", a non-reproductive, infertile, asexual, unattractive category.

The women in the survey talked of their changing bodies as being "undesirable" and therefore "invisible". They wanted men's sexual attention, and they also felt they would get more respect in general, for example for their accomplishments at work, if they still looked conventionally attractive (and research shows they are probably right!).

As we identify our negative images of an ageing or postmenopausal woman and understand where such images come from, we are in a better position to discard those images which are likely to hold us back. Of course there is a downside to ageing, but that is why we so badly need images of strength, endurance, humour, comradeship and wisdom to help us make the most of what may be as much as a third of our life.

Ageing is an opportunity to reinvent ourselves. The cessation of the grind of the workplace and the care of the family is at least in sight. Socially approved worldly ambitions are likely to be less compulsive. The menopause teaches us a certain bloody-mindedness, since we have less energy to waste and less inclination to do so. If men no longer find us pleasing, we no longer have to strive to be pleasing to them. We can simply be ourselves, and take time to find out who "myself" is.

You might try the following exercise: take a sheet of paper and a pen, and write "ageing" in a circle in the middle of the paper. Now write down all the words and phrases that pop into your mind, without forcing them, and which you associate with ageing. When you have finished, take a good look at what you have written. Which bits are positive, and which are negative? Can you trace why you have each association – does it come from your own personal or family experience, or is it a message that you have unconsciously picked up from society? Do you need to challenge any of your associations? Are you looking forward with confidence or fearfulness?

CHAPTER 2: Identity and Archetypes

A major problem for the ageing woman is: who am I as I get older? If I'm no longer what I was, what am I now? The danger is that we may fall into the trap of seeing ourself as a Victim. What we need to do is see ourself as a Hero. "Victim" and "Hero" are two ways of being, roles that we might play in a theatre, except that we play them in real life. They are both archetypes. The psychologist Carl Gustav Jung had much to say about archetypes.

Jung, originally a disciple of Sigmund Freud, stated that for every patient who came to see him who was in the second half of life (which he defined as 35 years and over), the important issue was essentially spiritual, in terms of a search for new meaning in life.

He said that young men and women must concentrate on love and work, but later in life they need to find neglected aspects of themselves, become a more complete individual and rediscover neglected ideals and values. They need to undergo a process of individuation, of developing themselves to be more truly who they really are. The search for the meaning of life and the answer to the question "Who am I?" are two sides of the same coin. A feeling of satisfaction with regard to both a sense of meaning and our identity is essential to mental health.

Jung studied world mythology, with its gods and goddesses, particular motifs, symbols and images, and was amazed to find that some of these symbols and images and motifs appeared in the dreams of his European patients, most of whom could not have encountered them personally. Jung concluded that these were universal symbols, found in what he called the collective unconscious, which is often talked about in a rather mystical way like some sort of spirit world.

Jung saw the collective unconscious originally as based in the physical structures of the brain, possessed by all of us as part of our inheritance as members of the species *Homo sapiens*. Our brains are constructed in such a way that they are predisposed to recognise certain symbols, to find constellations of meaning in certain plots of stories or particular characters. Jung referred to these universal symbols as archetypes.

Animals are very limited in their ability to use their brain to mull over various options and make decisions about what would be best to do in any circumstance. They cannot imagine themselves in different situations and outcomes and feel which would be the best result. So their behaviour is largely guided by instinct; their brain is wired to behave in a particular way, and they have very little freedom in that respect. The larger the brain of course, the more likely it is that their behaviour can be modified by learning.

As humans, we have great plasticity of behaviour and have the impression that we have free will in most things we do. We do have drives such as hunger and sex, but even then we can learn to ignore them. In the midst of this plasticity, the archetypes give us some sort of structure to help us negotiate life, particularly our social life. Jung argued that the archetypes organise our behaviour in a way analogous to instincts, although they are much less concrete and restricting than instincts.

Jung suggested that archetypes are predispositions to react in a certain way, but the detail of how we react depends on our own personal psychological makeup, including our personal unconscious and all the various neuroses we may have lurking there. The compulsivity of our response also partly depends on whether we recognise what is happening, in other words how conscious we are of the archetypes within us.

It is hard to prove the existence or otherwise of archetypes, beyond the most obvious, but we might find parallels in the way the brain has a preexisting ability to cope with language and the rules of grammar and syntax, while the particular form that language takes is dependent on the particular culture an individual develops within. Our brain also seems predisposed to organise lists and hierarchies of things, so that for example we can be easily able, at a young age, to recognise and name many different types of plant, and types of animal (or in modern times, types of cars!).

Jung suggests that there are as many archetypes as there are typical situations in life. Thus there are certain characters, and certain episodes in life, that strike a chord within us when we encounter them. Since they have a special significance, people have made up stories about them in every culture, in every time and place, since the Paleolithic (or Old Stone Age). As well as finding them in our stories, myths and fairy tales we can find them in our dreams, and in what Jung called active imagination – waking fantasies and visualisations.

Examples of archetypes that Jung wrote about include the Anima, or female principle within the male, and the Animus, the male principle within the female; the Wise Man; the Great Mother; and the Self, or centre of the whole psyche (whereas the ego is the centre only of the conscious mind), which in some ways corresponds to the divine within. Jung wrote of the dangers of inflation when archetypes are awakened. If our conscious mind – our ego – starts seeing itself as the literal embodiment of the Wise Man, or the Great Mother, or the Self, rather than understanding them as organising principles, there is a danger that we will end up thinking we are the Messiah for example, and lose our coherent sense of self.

Archetypal situations include the idea of rebirth, reincarnation, resurrection; the death of the old self and the birth of the new; transformation of our selves. This is a typical situation because our sense of self must change all the time as new events and relationships change the inner image of our self, but at the same time we need a sense of coherence and continuity. Any major change must be handled in such as way that it does not threaten our mental stability.

There are also archetypal symbols, for example the self may be represented as a child, or an animal, or an egg. A significant symbol is the mandala, which is any geometric figure such as a circle, a wheel, a square, or a cross; anything arranged spherically or radially or perhaps combining a circle and a square. It may be very elaborate, as in religious images, or simply appear in a dream as a square fountain in a circle of flower beds. The mandala is an ancient religious symbol, and Jung found his patients would dream of one when undergoing integration of the sense of self. Jung also found the old practice of alchemy to be symbolic of the self's journey to transformation and integration.

It is worth bearing in mind that Jung was a product of his particular culture, and his writings are far more sexist than would be the case if he were writing today. Jung's views were patriarchal and the typical person is usually assumed to be male (remember de Beauvoir's male Subject). Archetypes are generally discussed in terms of how they affect men, and a healthy woman is generally assumed to be a housewife. However, his ideas have been developed by women psychologists since his time, and huge numbers of archetypal symbols are listed and interpreted today. How many are truly archetypal is another matter.

Figures most likely to be archetypal are those that turn up both today and from earliest times. These include Mother and Father, Child,

Maiden and Young Man; Ruler/Leader, Hero, Warrior; Caregiver; Wise Man and Wise Woman, including medicine men, herbalists and midwives, shamans, priests and priestesses; the Joker and iconoclast; and the varied images of the integrated Self such as the Christ or other divine figures.

What evolutionary pressures would drive the natural selection of such a collective unconscious? We are social animals, and have a lengthy period of childhood dependence which allows the development of our relatively huge and complicated brain. In terms of family, this means we (ideally) have a mother who bears us and gives us primary care, and we also have a father who cares for both children and mother while she is burdened with such intense and lengthy care. The significance of the grandmother (which we shall be exploring later), and no doubt grandfather also, probably helped give rise to the wise woman and man.

Our hunter-gatherer ancestors may have lived in groups of about twenty or thirty, which gathered together in much larger groups for time limited periods. It would help maintain this useful social arrangement and reduce interpersonal conflict if there were archetypes for friendship, mentoring, leadership, caregiving, story telling, and the joker or fool who can defuse tension by making us all laugh.

Frightening situations such as hunting or warding off large animals, fighting between social groups over scarce resources, natural disasters, or decisions needing to be made when unexpected changes occur, might call for the leader, hero, warrior, shaman, peacemaker. The ability to invent and innovate and follow new paths, that has made humans so successful, would need a dynamic tension held between impulsive, confident young people, scornful of the old ways, and more conservative and cautious older people.

The ability to quickly learn and adapt to all sorts of social circumstances, to have potentiality wired in to our brains by our genetic makeup, would give a significant evolutionary advantage. A baby knows how to behave towards a mother to make her feel rewarded, just as a mother knows the nuances of how to treat a baby, infant or older child (if the archetype has been awakened by "good enough" mothering in her own childhood). Perhaps knowing how to respond to different situations, such as temporarily setting aside one's individuality to follow a leader in a tight spot, has been useful for our remote ancestors' survival.

It may be that the collective unconscious organised us just enough that we were able to respond "instinctively" to the complexities of life in a way that promoted survival, but without the complete lack of flexibility that goes with true animal instincts. There was plenty of room for learning and development, so that the archetypal content could change according to the person and culture it found itself within.

It was probably right from the start that the archetypes of the collective unconscious organised our inner world as well as our outer world, since this is equally important. Humans are unique in having a sense of self that can be manipulated in the mind so as to give us pictures of ourselves in every imaginable circumstance. This gives us enormous advantages in terms of planning and foresight, memory and learning, empathy and social interaction, but also requires a sense of stability and continuity since that image of the self is a construction in the brain rather than something concrete that we're born with.

The archetypes of the Self and mental integration are essential here. Even if the brain is a collection of different functions, the ego must have a sense that all is functioning as one smooth whole, and that the self I was yesterday is the same self that I am today. Even if I feel transformed, I must still feel that I am me and not somebody else. If a new experience happens that just doesn't seem to fit with my idea of who I am, some way must be found of digesting it. Sometimes the new aspect is just shoved out of the way into the shadow part of myself. The shadow is the part that is available to my consciousness if I don't repress it, but that I don't like and don't want to think about. Eventually however there must be some integration or the shadow will come out somehow, possibly in neurotic behaviour.

Jung lived through the Second World War and was very conscious of what devastation the complexes within us can cause if they are not recognised and dealt with in healthy ways. The shadow tends to get projected outwards onto other people, as a way of getting rid of it. Thus if I am mean, but don't like to think so, I will hide that fact from myself, as part of my shadow. To stop it bothering me, I will project it out on other people, saying I can't stand so-and-so because they are mean. Generally speaking, the things we hate in other people are the things we can't stand in ourselves. Jung believed that at the time of the Second World War, millions of people were encouraged to project their shadow onto the Jews with catastrophic results.

Jung also noted the harm that unintegrated archetypal forces could do. The Nazis let loose the archetypes of the Leader and the Warrior, and also fell prey to the *daimonium* or inner demon, the spirit in archetypal form which if awakened but unconscious can lead to the most terrible compulsive behaviour. Jung wrote at a time when the horrific possibilities of the atom bomb were clear to see, as the Cold War followed the Second World War. He felt it was essential that the forces of the unconscious were made conscious to defuse their power, and was angry and frustrated at how many people sneered at psychotherapy.

> "Only, heaven preserve us from psychology – *that* depravity might lead to self-knowledge! Rather let us have wars, for which somebody else is always to blame, nobody seeing that all the world is driven to do just what all the world flees from in terror." (Jung 1968 p253.)

Stereotypes are very different to archetypes. Stereotypes are cultural ideas of what certain people should be like, and often act as a straitjacket on those people. Stereotypes of older women for example may include ideas of foolishness, narrow-mindedness, gossiping and uselessness. Grandmothers are supposed to be easily shocked (so there is talk of a play you could safely take your grandmother to), and old wives' tales are ridiculous stories believed by gullible old women.

We need to reject the stereotypes that others try to foist on us and seek out archetypes which we can awaken within. The three we are most concerned with are the Hero, the Great Goddess and the Wise Woman. We will start with the Hero and her journey.

CHAPTER 3 – The Journey

It is in stories, myths and fairy tales that archetypal characters and situations are organised and displayed. These stories usually involve a journey of some kind, literal or metaphorical. There is a Hero in the story who has some sort of task to undertake or adventures to undergo, and at the end things are not the same as they were at the beginning.

These stories can be used as an explanation for the world we find ourselves in, so that things seem to make sense and to follow certain rules, and are not simply random and chaotic - creation myths are one example, supernatural explanations for natural phenomena such as thunder and floods are another.

However, universal stories are also about the development of the Hero – and the Hero can stand for the self. Humans have always told stories, whether sitting round the fire at night or writing books or creating movies. It may be that a major function of stories is as an aid to the universal problem of identity and meaning in a changing world. By awakening the archetypes within us, a story can help us to integrate new events, experiences, relationships or problems in such a way that we increase our ego strength rather than feeling confused or chaotic or descending into depression.

By entering the mythic dimension we move beyond the ego to the transpersonal realm, integrating conscious and unconscious into a new whole. We become deeper than we were, but at the same time less egotistical and "full of ourselves". It is likely that stories or episodes that particularly resonate with us have themes that are relevant to whatever situation we find ourselves in at the moment.

Joseph Campbell was a follower of Jung. In *The Hero with a Thousand Faces* he looked at myths from all over the world and different time periods and showed how they represented universal and timeless human themes about the development of identity and meaning. He analysed the elements making up the archetypal Journey found in myths and produced a universal narrative structure - the component parts of a story - along with an explanation for its power.

The journey of the hero is about a continual dying and rebirth – death to the old and birth of the new. As we enter new phases of our life, particularly when we undergo an abrupt but far-reaching changeover such as the menopause or a serious illness, the need for the death of the old self and a rebirth of a new self become particularly important. A story of a hero's journey can help us through that transition. Again it is worth bearing in mind that Campbell, like Jung, had mainly a male audience in mind; but his ideas are easily applicable to women.

In summary, the first stage of the journey is separation, or departure – the call to adventure, the supernatural assistance and the passage into the realm of night. Then there is the stage of the trials and victories of the journey – facing the terrible dangers and winning the prize. Finally there is the return and reintegration into society, when the prize or knowledge or wisdom gained is put at the disposal of society.

Campbell sums it up as:

> "Separation - initiation - return… A hero ventures forth from the world of common day into a region of supernatural wonder: fabulous forces are there encountered and a decisive victory is won: the hero comes back from this mysterious adventure with the power to bestow boons on his fellow men." (Campbell, 1949 p243)

The Hero is the archetype of the individual ego, which must adjust to the constant flux of life and die to the old and be reborn to the new. We might look at it as the death of the old identity of being in the second third of life – the full moon, the productive mother archetype (which may not literally have given birth), and taking on the new identity of the ageing/post-menopausal woman, drawing on archetypal aspects such as a crone goddess, a grandmother, a wise woman, whatever feels right for us.

The Journey could happen over and over as each new adjustment has to be made – perhaps a frailer body, chronic health conditions such as IBS or arthritis; the care of elderly parents who once cared for us, or becoming an orphan; the changing relationships with children who are themselves ageing; new roles in family and career, altered material circumstances, changing interests, changing friendships, changing sexuality, changing spirituality, and the facing of our mortality.

First comes the call to adventure, which tends to come in supernatural ways in myths. For us it is ageing itself and therefore inevitable. The call can be refused or denied, usually with dire consequences until the Hero is disastrously forced into the journey. With women it is harder to deny the call because of the menopause. We are thus more likely to be propelled into the journey to wisdom which is of advantage to society as a whole (the 'boon') even if unpleasant for us at times. Possibly the continued use of HRT (hormone replacement therapy) could make a difference. Where the call to journey is denied, death and rebirth cannot take place so stagnation results, which may even lead to depression. Advertising is one pressure to deny the call – to stay young and beautiful, have the botox, and pretend that nothing is changing.

When the Hero sets off on the journey, she is usually accompanied by a protective figure, often a wise man or wise woman (think of Merlin, the Fairy Godmother, Gandalf and Galadriel from *Lord of the Rings*). The Hero is protected by ageless guardians; Mother Nature herself supports our quest, giving us the feeling (if we respond to the call) that all is as it should be.

Then we come to the threshold guardians on the edge of the darkness, the unknown, beyond the protection of society and the everyday known world. In other words we embark on a journey that only we can undertake; our nearest and dearest can't travel with us even if they encourage us on our way. It feels dangerous to cross this threshold, but if we have courage enough to press on then we will come through OK. The feeling of danger is the pain we feel when we face up to the challenge instead of repressing everything. That's why the archetype is a hero: it is ultimately courage that wins the prize. The threshold is also about entering a psychic realm, rather than a material one, that is beyond the normal five senses of the everyday world. This can give the journey its strange and terrifying aspect, for we are usually far more at home in the outside world than the inside.

We now reach "the belly of the whale" – we have travelled into a kind of self annihilation. To the outside world the Hero may have appeared to have died, or disappeared (withdrawn). In psychic terms we go inward, to be later reborn. Passing the threshold guardians means we have proved ourselves capable of going within and undergoing metamorphosis. This stage is also described as being within the temple. In myth the Hero may be completely torn to pieces at this point, and it may feel like it psychically in real life, facing the death of the old self.

We enter the initiation stage with a succession of trials, tests or ordeals to undergo. We may be aided by helpers, companions on the way. What trials might there be for us? It may be the burden of extra demands on us just when feel most fragile and menopausal; perhaps our marriage becomes very rocky; perhaps we have to face up to the end of our career. Our helper and guide may be a counsellor, or a priest, or a friend who has been through it herself, or a support group of peers. We may feel we are facing it entirely alone, in which case it is important to remember that we are not on our own, that this is a universal female journey. Women have undergone such trials before and will again. Celebrate them in your mind, and have compassion for those still undergoing the journey. In any case remember the protective figure of the Wise Woman.

Next we come to the meeting with the Goddess – the creator of the world, the source of all birth; also the death of everything that dies. This is about the unity of the opposites and may be symbolised by some kind of marriage. In fairy stories it may be about embracing and loving an ugly woman, not being put off by the face of the hag – when the hero kisses her, the hag turns into beautiful woman. Or for women, the story of Beauty and the Beast may be helpful – the Beast when loved turns into beautiful young man.

Embracing the opposites means holding the great paradoxes in our mind – the fact that life consists also of death, of dark as well as light, of suffering as well as pleasure, of shadow as well as persona. If we can hold the doubts and questions in our mind and not reject them – why is there death, why suffering – even if we don't find an answer that we can put into words we can come to some sort of resolution, so that we can go on in a hard world and see its beauty.

Campbell next talks of meeting woman the Temptress, which represents temptations that may lead the hero to abandoning the quest. This is particularly about physical or material temptations, perhaps to turn to comfort eating or drinking or "retail therapy".

Next comes Atonement with the Father, in many ways the climax of the journey. Having been tested by various trials, the hero now meets whatever is most powerful in his or her life, and is in some sense killed by it in an act of initiation. This may be a male figure, or may not. Even if it is one's father, it is the archetype that we confront and the death we undergo is also archetypal or symbolic. This stage is commonly encountered in initiation rites as the ultimate challenge, or ultimate terror, the climax when the initiate is changed for ever. In a fairy story, it may be a fearsome ogre or dragon. In the myth of Inanna, Sumerian Queen of Heaven, which we shall explore in Chapter 14, it was her sister Ereshkigal, Goddess of the Underworld.

For us, it may be the moment when we realise that death is real and unavoidable, our mental defences stripped away; the moment when we realise that the loved one who holds our happiness in his or her hands no longer cares and the relationship is over; the moment when all that seemed dear and familiar to us, whatever we gladly assigned as our ruling star, to have power over us, has been stripped away or betrayed us. This is the moment when as in the myth of Inanna, we are slain by Ereshkigal and hung on her meathook for three days.

The time of terror and despair is followed by apotheosis – to knowing the divine. Having died to the self we are open to the spirit. We move beyond the opposites to divine knowledge, love or compassion. Having moved beyond the ego we enter the transpersonal realm; we are in touch with unity consciousness. Since all we held dear is dead anyway, we have stopped grasping and have let go; paradoxically at that moment, we are reborn, renewed, at peace.

How long the process of death takes, freeing us to this rebirth, will depend entirely on individual circumstances. It will no doubt take longer than you would like. It may be a little death or something truly significant. Death may come all at once or in a series of blows. But we have to die first in order to be reborn. By dying to the familiar we break down old boundaries and suspend judgement. We find that what we thought were hags are goddesses, and dragons give up their store of gold.

By undergoing this initiation we find the ultimate boon. In myth and fairy stories, this may be a token such as a magic crown, or ring, or wealth, or knowledge, or new powers. It may be the elixir of life or the Holy Grail. For us it may be an expansion of consciousness, a new wisdom, compassion, understanding, new meaning and purpose. Whatever it is, it is the goal of the journey and is needed not only by the Hero but also by his or her community. This goes beyond the individual.

The third phase for Campbell was the Return. This is not just an afterthought, but may be as dangerous as the departure or the initiation. The Hero may now carry wounds or scars that make the return even more difficult. The more significant the boon, the more the difficult it may be to make the escape from the monsters guarding it (perhaps being the fears we have previously repressed and which now become conscious). There is also the danger that the Hero will refuse the return, and remain withdrawn rather than bringing the boon back to her community. Once again the Hero may need helpers to return, companions or mentor.

The Hero crosses the final threshold back to the "real" world, at the end of her psychic journey. This is the stage of resurrection, and the task now is to integrate back into the mundane world after all the adventures. How can the boon be shared with the Hero's community, which has not undergone such a quest, may not realise that such a quest exists, may have no understanding of the boon's value and may even be frightened of how we have changed? This is often the most difficult part of all. For the older woman it may mean simply living with her new wisdom, acceptance, enthusiasm, way of relating, social message or whatever it may be, without trying too hard to convince others of its value. Those with eyes to see and ears to hear will quietly absorb the message.

The Hero has now achieved a balance between the material and the spiritual worlds. In myth, this may be represented by a divine figure who can move between the two, such as the Christ or the Buddha. Most of us will simply be an anonymous hero, having learned to give up attachment to our idea of who we are and to let change happen. The archetypes we identify with here tend to be sages and hermits, or even Christ – the symbol of the Self, of the fully realised and individuated (Jung) or actualised (Abraham Maslow) self. Ideally we recognise the persona (the face we present to the world) as a role that we take up and play with, rather than our real Self. The more we can achieve this, the more easily we can cope with future changes to our sense of identity.

Campbell talks of the new freedom to live – "the hero is the champion of things becoming, not of things become, because (she) *is*" (Campbell, 1949 p243). The Hero is not stuck in the past, bound by memories of what was, equating what has been with what ought to be, or fearful of death and the future; not fearful of time "destroying the permanent with its change". The Hero can live in the moment. With resistance to change overcome, life flows. And the Hero's freedom to live allows life all around to flow.

Campbell allows that there are many variations on the basic theme outlined here – the elements may not all be present, and the journey may not be a once and for all. The timescale of the journey is always different depending on the person and circumstances. One may go round many times, like a spiral. After the death of the old self and the birth of the new, we may start fall back into regret and need to go round again, perhaps many times until the birth process is complete. And later on, when the new self may have become hardened and change once again is difficult, the process has to start all over again.

Campbell's work struck a chord with many people and has inspired a number of fascinating books on the mythic journeys that we undergo, or the personal meaning we can extract by examining certain popular myths and fairy stories. One modern development is the application of his schema to film, George Lucas' *Star Wars* being a popular example. The award-winning Hollywood story consultant, Christopher Vogler, who has been a consultant on many Disney productions including *The Lion King*, developed his own framework from Campbell's work outlined in his book *The Writer's Journey*. You may find it interesting to compare the two.

You may even want to see if you can turn your own life story, or recent episodes, into a mythic journey with yourself as Hero. Although this can be a useful exercise, we can also choose some archetypal journey or metaphor that resonates for us and helps us to identify ourselves as Hero.

CHAPTER 4 - Some Journeys, Stories and Metaphors

Sometimes it is helpful to explore one particular myth or story in detail and apply it to one's own life. Sometimes just one episode or even one image rather than a whole story can be useful. It may well be all you need, since the short piece of story that really resonates for you may be precisely what you need to feel integrated at that period of your life. The important thing is to see yourself as the hero of a timeless, universal journey: the journey to a new, more truthful version of yourself. The story or metaphor may be about your life as a whole, or one aspect of it, or about yourself, or your hopes or projects.

Time cycles fit in well with the issues of ageing. We might see our life in terms of a day, the cycle of the earth; from the morning to the afternoon, evening and finally the passing away of the light into night time before dawn returns the next day. We have seen how Jung talked of the tasks of the morning (love and work) and the tasks of the afternoon of life – to fully individuate, or become all that you are. This is the time to concentrate on yourself, even though you continue to love those around you.

Death is often seen as night, and Dylan Thomas told his father to "Rage, rage against the dying of the light". One of our tasks is to make friends with that dying light, though while we are in the afternoon nighttime is still a long way off, and evenings can be very beautiful.

Seeing the day as an episode of our life, we can use the metaphor of the days cycling round. Thus we have proverbs such as "It's always darkest just before dawn", and concepts such as the "dark night of the soul". Within the day we may see a cycling through times of sunshine to rain and back to sunshine again, knowing that the rainy times are necessary for the fertility of the soil.

The cycles of the moon are particularly suitable for women, since we have long been associated with its cycle due to our own menstrual cycle. As we have seen, there are images of the waxing moon (maiden), full moon (mother) and waning moon (crone). As well as standing for the complete life cycle, they may be taken represent particular episodes or projects, with beginnings, growth, maturity, followed by a time of falling away and a rest period when new projects begin to take shape. You may want to honour the cycles of the moon by practicing some rituals or magic, such as found in DJ Conway's *Celtic*

Magic (you don't have to believe it literally!) Traditionally, when you wanted something to come to you, you worked the spell during the waxing moon (that can be cupped in your right hand), and when you wanted something to diminish you worked the spell during the waning moon (when the moon can be cupped in your left hand).

Another cycle of time is the seasons, the cycle of the sun. It can be too easy to overlook the cycling of the seasons in the modern industrial world, where we are insulated from the natural world by our towns and cities. For most of history, and for many of the people in other parts of the world today, life depends directly on the fertility of the earth, whether natural or farmed. Agriculture demands that we know the season for planting, flooding, harvesting etc, and that the deities send the right weather at the right time.

Old myths explained, and through ritual helped to ensure, the death of fertility at the start of winter, the long fallow period followed by the fresh regrowth of spring climaxing in the harvesting of fruit and seed, before falling fallow once more. It seems likely that blood was sometimes spilled to ensure this cycle continued, in particular so that the dreadful time of late winter when it seems that spring will never come again, is in fact followed by the shoots appearing in the soil.

We can use these metaphors of time cycling both for our own life spans and as metaphors for good times cycling through to bad times and then to good times again; seasons of hope, freshness and planning giving way to productive times, then times when it all seems to fall away and lie fallow until ideas or optimism begin to stir again. Sometimes we may feel that we are in the myth of the old king who was killed and his blood spilled on the cut fields to ensure next year's crops; then we are reborn as the goddess's son with the coming of spring.

We can also look at the old religious festivals (Christian or pagan) that celebrated the cycling of the seasons. Often there were particular gods and goddesses attached to them, and particular myths. The Celtic tradition offers us the cross quarter days, the four festivals of Imbolc (the very beginnings of spring on 1st February); Beltane (May Day, when the cattle were driven between the purifying fires to bring luck and fertility); Lughnasadh or Lammastide (August 1st, the beginning of harvests); and Samhain (pronounced Sowan – 1st November, the end of the growing year and the start of winter).

Samhain may be of particular interest to some crones, as it corresponds to the waning moon, and is a time when the threshold between this world and the spiritual world was thought to be very thin. Hallowe'en, the night leading into Samhain, is traditionally the time to see ghoulies and ghosties. This is the time of the wise crone goddess, also the fearful aspect of the goddess – the keeper of the mysteries of death. This may be a time to look back, to reassess, and then to let go of the old so that there may eventually be the birth of the new.

Imbolc comes at a time when winter still seems to be in control and there seems little sign of the sunshine to come. Imbolc reminds us that the signs of spring and renewal are there, in the snowdrops and green shoots poking through the earth. We can take it as a lovely reminder that our own personal winter can end when we least expect it and the promise of new life, new projects, new relationships and new beginnings is here before us.

Then there are life cycles. There is the life cycle of the plant – the seed or acorn lying quietly in the ground, with no sign of life; then the first shoot appearing above the soil. Remember that the seed has to "die" in order to germinate. We then have growth, with new roots reaching down to become firmly set in the soil, giving stability and sources of nutrients and water; before that happens the seedling is very vulnerable to unkind winds and weather. You may be in a new situation that requires you to set down new roots. At the same time growth appears above the soil, reaching up towards the sun. The plant eventually flourishes and blossoms, but not before the right season. Then, as always, there is a falling away, a fallow period. The plant now appears dead, but either the roots of the perennial are waiting for spring, or the annual has scattered its seeds to blossom in their turn next year.

There are many different kinds of plant. You might compare yourself to a mighty oak, or a tree on a windy heath, gnarled and bent from years of storms but strongly rooted and holding on. You may be fruitful, or have produced showy blossoms; you may be a little scarlet pimpernel, beautiful but hidden from view. Many plants had religious significance in old myths, and you may be able to look up one you identify with and see what is known – was it associated with healing, or the underworld, or powerful magic, or a particular divinity?

Animal life cycles might include the butterfly, with the cycle from egg to caterpillar to chrysalis to butterfly to egg again. The image of the crysalis can be particularly useful in times of great change such as the menopause. Within the hard outer casing of the cocoon the inner creature turns to a mushy soup of formlessness; gradually a new organisation starts to take shape; eventually the creature will be reborn as something completely different to its former existence. If it could think, what on earth would it feel like to the caterpillar to experience itself being completely broken down, to be reborn with a new beauty, and with the new power of flight? The crysalid stage can last for ages; even when first emerging, the wings are still flimsy and weak and unusable; at first it must feel very fragile and vulnerable, perhaps as though it wasn't such a good idea in the first place. But then blood is pumped into the wings, energy comes back and the butterfly can take flight.

The frog has long been used as a symbol of metamorphosis, with the change from tadpole to adult frog, from fish-like creature in the water to four-legged animal that more closely resembles ourselves. Just as the caterpillar moves from being earth-bound to taking to the air, the frog starts as a water creature and then can move to the earth. As they move to completely different elements, so might we completely change the milieu in which we move.

As the snake grows, it has to slough off its old skin, which is now too small, too restricting, and reveal the new skin below. The shedding of the old skin can stand for the stripping off of old illusions, about ourselves or about other people. It can stand for the dying of the old self and the transformation to the new. An age-old symbol of wisdom, it can represent our ability to adjust to change and take on new roles, new knowledge, new wisdom.

Hermit crabs have no shell of their own, but borrow an empty shell left by another crab that seems to fit them and take it as their home. Eventually they reach the point where they are now too big for the shell, and they have to leave it and find another. In that period of time when they are out of one shell and not yet in the other they are very vulnerable to predators and to dessication. Once they are in the new shell, there is a further period of comfortable stability and growth until the process of moving on must happen again.

We could compare these episodes to times when we have to move on, leaving the safety of home to go to University or a new job, becoming a parent for the first time, deciding to leave our partner, or facing retirement, perhaps moving into a care home when too frail to manage at home. These are all times when we move from the safety and security of the known, and risk the dangers of the world outside our familiar routine.

Even worse, the shell may get broken in a storm or smashed by a careless foot. We may feel like this when we are struck down by illness, or our spouse announces he wants a divorce, or we are made redundant, or we are bereaved. It may simply be the menopause itself, or the realisation of how fast we are ageing. We may feel completely lost, not knowing who we are anymore, all the old certainties gone. It may take a long time to find a new shell, a new stability, a new persona, and we are very vulnerable to the elements while in that time of transition.

There are some boons in being taken out of our security zone. In particular it is when we are out of our shell that we are most open to intimacy. Most of the time we hide away behind the shell of our persona, the idealised picture we try to present to the world of who we think we ought to be, rather than who we really are. When we lose the sense of our persona, we are most likely to enable others to see the real self. This is terrifying, because of the feeling that if you knew me as I really am, you wouldn't like me anymore, might even be repulsed. But you can't have real intimacy with a shell, only with the vulnerable creature inside it.

It is possible that some people won't like to see us as we really are. If they are used to leaning on us for example, they won't like to see that we can be as weak as anyone else. If they like to share the belief that we are victims, they won't like us being strong. But pretending to be something we are not in order to keep those around us comfortable is a sham. Whatever we are, we will find people who are sufficiently authentic themselves that they can love us for who we really are. It may take time to find them, and that may be part of the search for a new home, but we will find them.

We cannot get in touch with the core of the divine inside us unless we strip away the illusions about ourselves and our world that clutter up our ego. As we have seen in Campbell's schema, it can be when we feel most destroyed – when our persona has been shattered – that we come to a new sense of the Ultimate that is beyond words and beyond the opposites. On the journey, the times when we feel most unsafe and least sure of where we are going can be the times when we learn the most – when we are most open to spirituality and to one another.

However, there are very few of us who can live in a state of naked vulnerability all the time. There is no shame in moving on, indeed it is part of the journey, and we need to be able to comfort ourselves with the thought that others have been here before us, and we all find a new shell in the end. Until the next time.

Religious myths, as Jung found, are a rich source of meaning for life's journey. You might want to browse through the many collections of myths of different cultures (Greek, Roman, Nordic, Celtic, Native American, African, Indian and others). You don't have to be a Christian, for example, to find Biblical stories rich in metaphors as meaningful today as when they were first written. Ideas of exile, exodus, desert wanderings and other journeys, wounds and healing, death and resurrection can all find resonance in our own lives.

Books, poems, and sagas can all give us metaphors for our own journey. Some examples are the Greek *Odyssey,* the tale of the hero's long journey back from the siege of Troy to his home and family; John Bunyan's *Pilgrim's Progress*, a Christian parable of the journey through life and from which we get, for example, the phrases "Slough of Despond" and "Vanity Fair"; Dante's *Divine Comedy*, his allegorical journey from the "dark wood" he encountered in mid-life through to hell and then purgatory and on to paradise; and the story of Parsifal and the Holy Grail. Modern day stories might include Frodo's quest in the *Lord of the Rings*.

Which myth or fairy tale or story or film rings a bell for you, in terms of picturing yourself on the archetypal Journey? It may be one outlined here, or one you have come across yourself. You will probably find that different stories tend to fit at different times of your life. How did/does the story relate to your life? What did the various elements represent for you? Did you find it helpful, or did it simply feel right?

We will finish here with a brief look at symbolic objects. As we saw with Jung's mandalas, objects as well as characters can have archetypal significance. Symbols are things that carry meaning beyond what they signify, an obvious example being the Christian cross, which was a form of Roman execution but now symbolises death and resurrection. Symbols may not necessarily be of archetypal significance, in that they tend to be culturally specific, but since they encode much meaning in one simple package it seems appropriate to look at them here.

Symbols can be helpful as an organising or integrating function for minds under stress. I remember when my mother was dying in hospital from a stroke, I held a cross in my hand during the long night I sat by her bedside. Even though I had not yet made the decision to return to the Christian church and have never believed in heaven or miracles, I found it a comfort.

Tarot cards can also be used for their archetypal meaning, and are usually available in bookshops. They can be set out in a normal "fortune telling" spread, but used to organise our thoughts as to where we are at this moment in our life's journey: what has led up to this, what is the task that faces us, and our assets and problems, and some thought of what we hope or expect for the future. There are many books on the Tarot, including in the library, which explain the symbolism of the various cards and give suggestions as to how to lay them out.

As we have seen, animals or plant material such as seeds or blossom can be symbols of growth and rebirth. A woman might symbolise the sense of awe towards something greater than her small ego by bringing symbols of Mother Nature into her home: logs, pine cones, stones, fossils, driftwood, shells, or lighting candles in a sacred space – perhaps decorating it with coloured glass and feathers.

People often gather small possessions around them that may not be worth anything but which are very precious to them – a gift from a loved one; a child's first shoe; a lock of hair; a souvenir from a significant time and place. We may have symbols of our achievements: something to remind us of when we were proud of ourselves, or to remind us of how far we have come.

I have a two objects that are significant for me with regard to the menopause; one is a jade owl – associated with the crone goddess Hecate, and also symbolising the wisdom which I hope I am storing up as I age! The other is a mourning ring belonging to my great-great aunt Louie, who came from India. She seemed a rather more exotic, interesting and adventurous character than most of my relatives (and long lived), so her ring seems a suitable symbol of a post-menopausal Hero for me.

You may have a symbol of your own, or be drawn to one mentioned in connection with a goddess. If not, see if any object in or out of doors or perhaps spotted in a second hand shop draws your attention. Otherwise you may wish to do a visualisation exercise, meeting your Guide in a beautiful, sacred landscape and asking to be shown your own symbol for your journey.

For such an exercise, get yourself in a comfortable place where you won't be disturbed. Close your eyes (after you have read what you need to). Imagine yourself in a place that is safe and beautiful – it might be somewhere you know, or one you make up.

Now imagine that someone or something appears that is a reflection of the deepest and wisest part of you – your Guide. It might be a traditional shamanic figure, male or female, a Wise Man or Wise Woman, or perhaps it is some kind of animal. Just relax and see what, if anything, comes into your mind. When you feel it is there with you, you can explain your need and ask to be shown a symbol for your journey. You may find this is a peaceful place to be and you want to stay there for a while, possibly having some conversation with your Guide. If you find this kind of exercise doesn't suit you and no images appear, it doesn't matter; just sit quietly.

Afterwards, allow yourself to come back peacefully into the room, open your eyes and then ground yourself by looking around the room and noting what you see, and moving your limbs and stretching or wiggling your toes.

CHAPTER 5 – Meaning, Story and Sense of Self

In this chapter we explore why it is that seeing ourselves as the Hero of our own archetypal Journey can make a difference. All the journeys we have looked at – whether the standard journey of myths and legends as described by Joseph Campbell, or the briefer vignettes of looking at the turn of the seasons or life cycles – are stories that are used to express deep meaning. A story is truly archetypal when it can embody truths about changes in identity, whether in individuals or wider society.

If we have an age-related health problem we might persist in feeling like a victim, or we might accept that this is what happens as we get older and thank heaven we're not as bad as some we know. In the first case we are telling ourselves a particular story that says we have been picked on by vindictive fate and struck down in a meaningless fashion. In the second case our story is one of timeless truths in which we follow the path that was set out for us from birth and which our sisters have trodden before us. In surrendering to the unalterable ways of the universe we find some sort of meaning and thus the path is easier to bear, even if still a hard road to tread.

The importance of a sense of meaning tends to be recognised more when life seems meaningless, for we don't think about it very much otherwise. It is not necessarily about something concrete. If asked what gives their life meaning, someone might reply that it's their family, or their job, or their religion. At another time if suffering from depression, they might still have their job or family or religion but unaccountably feel that life has lost its meaning.

Meaning is a feeling to be sensed rather than about logic or intellect, thus it is ultimately about emotions. When someone is depressed their feelings are damped down, in order to avoid feeling emotions that the mind fears would be unbearably unpleasant. Unfortunately it is hard to clamp down on one set of emotions without affecting all of them. Thus a depressed person is unable to feel joy as well as anger or fear or grief, and often they are also unable to feel that their life has meaning. Everything just feels grey and pointless.

Meaning is related to the expression 'gut feeling'. We just know if we feel right and we know if we feel wrong, even if we can't explain why. The brain is often compared to a computer, and this can be a useful analogy in the complicated task of trying to understand how our brains work. However, the drawback to this analogy is that it can put all the emphasis on analytical thinking when that is only one aspect of the brain.

Neurologists such as Antonio Damasio and Joseph LeDoux have been an important counter to the computer model and have emphasised how basic and important emotion is to the functioning of the brain. After all, most of the brain's evolutionary history did not involve thinking.

To give one example of the centrality of feeling, people who are brain damaged by accident or disease in such a way that they cannot feel their emotions, are unable to make decisions or else make foolish ones. They can come up with all the intellectual data as to what will happen if they follow that path and what will happen if they follow the other, but they lack the 'gut feeling' that tells them which path to take. Without the gut feeling, the data is just a mass of data. It doesn't tell them what to do.

The existential therapist Irvin Yalom talks about the existential dilemma – the search for meaning and certainty in a universe that has neither. If everything will come to an end, if everything is due to the chance play of atoms, if we construct our lives out of our own heads, what meaning is there in life? Yet we must find meaning, for that is how we are made.

Viktor Frankl was a psychiatrist who was imprisoned in Auschwitz in the Second World War. He, and others, endured horrors that they would previously have thought unendurable, while millions died. Frankl observed that those who died, often died because they gave up – they lost some inner hold on themselves. He also observed that this was due to losing any sense of meaning or purpose in life, any reason to go on living. Those who could feel that there was some sense, or aim, or purpose in life, could carry on.

Frankl expressed it as learning that "it did not really matter what we expected from life, but rather what life expected from us." Rather than finding an absolute meaning that life had for all, each person had to find their own right action and right conduct. Even though they were all in the same concentration camp, each person had to find the answer that was right for them alone, in their own particular situation within that camp, and carry out the allotted task. That was their meaning. It might be to accept suffering as one's task; to die well; to comfort others; to survive as a witness to the camps, to let the world know what had happened; to survive so that one could ultimately be reunited with loved ones; to survive so as to seek revenge when that became possible.

Frankl concluded that we have a *will* to meaning, and that each person must find their own meaning for their own life. Meaning is something that people are able to live and die for. He quoted Nietzsche: "He who has a *why* to live for can bear almost any *how*."

What the people in the camp had done was respond to external events. No-one had chosen to be there. What they did have some control over was how they responded. When they simply saw themselves as victims – hard not to do under such terrible circumstances – they were much more likely to die from starvation or disease or suicide. When they saw themselves as the Hero in their own journey, whatever that journey involved, they found the will to survive through the meaning it gave them. It didn't matter if they had to be physically passive. What really counted was the attitude of mind.

The human brain is disposed towards making patterns out of chaos, seeing a familiar image in the unfamiliar. When we see a cloud, we can easily see a face or castle or ice cream cone. We see spooky faces in dark bushes. Likewise, we order chaotic events into a story. If we did not have some sort of organisation of the continuous stream of sensations battering on our brain all day long, we could never cope.

As well as the details of the events we experience, we also have the question of their meaning and their relationship to each other. We don't hold warm skin and milky taste and full stomach and sense of safety all separated in our infant mind, we have the one experience of breast feeding, or perhaps the experience of mother. Later as things become more complicated, we hold the whole thing together through an ongoing story, or narrative. This is particularly important to maintain the sense of self. We can't construct an increasingly detailed image of who we are unless the person to whom things are happening today is the same person to whom things happened yesterday.

For that we need a sense of permanence in a world in which everything is continually changing. Of course parents tend instinctively to give a child as much of a routine as possible, sensing that the younger they are, the less disruption they can take. Even so, what happens today will never be the same as what happened yesterday so neither can I be the same as I was yesterday. If who I am is built up by my experiences of what happens to me and how I act, then something must give a sense of continuity. That happens by telling ourselves the story of our life as we go along.

In this story everything needs to make some sense. There must be reasons for events in order for them to follow on coherently. Thus our brain tends to assign reasons and motives where perhaps none exists. I might blame some adverse event on myself because at least that gives me a sense of meaning and some control over events. Gods and goddesses are useful as explanations for happenings we don't understand.

So, instead of experiencing a series of random disconnected events hurled at us out of the chaos of an uncaring universe, our brain is structured to turn our life into a coherent story of which we are the hero. Story, or narrative, is the great meaning-maker that enables us to string a series of events together in a logical sequence, making sense of who we are and what is happening to us, making our life and thus our self seem continuous.

When children suffer serious discontinuities in their lives – perhaps through being sent from inadequate chaotic parents to live in a series of foster homes – they may find it hard to make a coherent story out of their lives, and the result can be a poorer sense of identity, a more fragile ego and therefore more fragile mental health. Social workers are encouraged to construct life story books for such children, and to work with the child to make it truly his or her book, with facts, names and pictures clearly and simply woven into a tale that makes sense from birth to present day. The child is helped to turn their life into a coherent and meaningful story and thus have a stable sense of who they are.

Counselling can be helpful for us all in constructing a coherent story of our lives. As we tell our story to the counsellor we impose sense and meaning on it, or find sense and meaning through our interactions with the counsellor.

Person-centred counselling originated with Carl Rogers, who laid down the three core conditions of the therapeutic relationship: acceptance, genuineness and empathy. When clients themselves are asked what is helpful they have also stressed the value of the chance to talk – to give new meaning to their experience. By being able to put their thoughts and feelings into words, they are able to make sense of past and present experiences and to control the meaning of those experiences. They are able to restructure their thoughts and feelings so that their future is more hopeful, and they see new meaning and order in their life. It is often by speaking out loud and hearing what we say that we first come to understand who we are.

We sometimes feel we only understand what happened as we describe what happened, as we listen to the words coming out of our mouth. It can even be as though we are getting to know someone – our own self – for the first time. This may well be because that understanding, or meaning, or aspect of our self didn't actually exist until we had organised it into words. Thus through language – telling our story – we can find meaning.

Story, or narrative, is a step up from description – it is the telling of a series of events rather than just one. It is so basic to the way we organise meaning in our internal world that we are rarely aware that we are doing it. In fact David Howe suggests that what we do not structure as a story simply doesn't get remembered.

Our sense of self, our identity, is both shaped by the story we tell ourselves and essential for creating the hero at the centre of the story. Our consciousness or thinking brain enables us to imagine ourselves in a particular situation (past or present), so that we have foresight and planning, as well as learning from our memories. We can also imagine ourselves in someone else's situation, so we have empathy and social skills. To do these things, we have to have an idea of a 'self' that we can use to imagine with. Where does this sense of self originally come from?

Damasio suggests that this arises originally from feelings and sensations within our body. As we first experience life, we sense what is happening and because our body reacts to it (with good or bad feelings), we also get the feeling that it happens to *me*. As experiences build up, so does that sense of me. Thus we have the sense of self, a sense of who we are.

We may note that instead of one event of knowing, we now have a split between the known and the knower. We feel we have a little person in our head who is experiencing, knowing, deciding. In a way the world is experienced at one remove. This is essential for human consciousness, but forever separates us from the world.

Our sense of self is very fragile to begin with, and needs a lot of empathic care from those around us (usually our parents, particularly our mother). Our sense of who we are arises from our relationship to the outside world, and if that suddenly lets us down, then we are not who we thought we were and there is a disruption in our sense of self.

Child psychiatrist DW Winnicott came up with the phrases "the ordinary devoted mother" and "good enough" mothering to describe how generally speaking, most of us manage to grow up without anything going badly wrong. However, any of us may present a false, supposedly more acceptable self to the world (our persona), while keeping our real self with its shadowy, less acceptable aspects, more or less well hidden. Since we are social animals, it can be dangerous for us to go against the expectations of those caring for us and of our community.

Another source of our sense of self is in what we *do*, how we *react* as well as how we experience. There is a feeling of authorship of our actions – because our feelings are all body-based, we 'resonate' with what we do, whereas we only 'notice' what happens. As our brain tells us we have willed an action, our body tells us we have completed it. Thus we know who we are, what we can and can't do, and what we do well, and the picture we continue to build up from our sensations continues to develop as our sense of self. That sense of authorship is essential for good mental health.

We can only be aware of our conscious self, although experiments show that most of our decisions are initially taken unconsciously. Our conscious self is actually quite limited in its awareness. With most of the responses we make, thousands of them every day, we hardly even notice that we are making them. We can do many things on autopilot and it doesn't usually affect our efficiency.

Sometimes we even work towards that feeling of loss of conscious authorship, for example in sports. Players work towards getting so practiced and skillful that they have a feeling that "the game played me" rather than "I played the game". Loss of self consciousness may also sought in the creative process, by losing oneself in what one is doing.

Our sense of self can be manipulated in our brains in all sorts of ways. As we imagine this "I" in different situations, practical and social, we can work out what would be likely to happen if we did this or that. Thus we can plan for the future. Or we can act out our interpretation of what has happened to us in the past and is happening to us now, including our interactions with other people. Thus we impose our meaning on these events.

In order to run this internal theatre, where "I" join imaginary actors in dreamed-up scenarios, our brain's frontal lobes needed to evolve inhibition on a large scale. We can, for example, imagine or remember a dangerous situation even to the point that our adrenaline starts flowing, but not get up and run. Our impulses have been decoupled. We can roam over the whole universe in our heads without moving a muscle. We can focus in on one particular aspect of a problem in a state of deep concentration, excluding all irrelevant data. Or, we can let our minds wander – lateral thinking – and allow all sorts of random associations enter consciousness, so that entirely novel juxtapositions of images and ideas occur to us. This is the basis of creativity.

As well as imagining and planning the future, we can also learn from the past. We can rerun scenes from the past in our internal theatre, and think about what we might have done (and therefore what we might do in a similar situation in the future) in order for the outcome to be improved. This can be quite unpleasant when we become stuck in a loop – when we keep compulsively going over and over the same memory, thinking "Oh God, why did I do that?" or "I should have said this, or that." This is particularly likely when what happened seems to have threatened either our physical self or our sense of identity, our sense of who we are.

As well as accurately depicting who we are, we can also pretend to be someone else. This too is essential, in that when forming an image of how we will act, we also need to guess fairly accurately how other people will react. Will they support us or not? Will they be happy or angry? Will they understand what's going on, and will they see it from my point of view? We begin to understand that other people may think differently, and that all minds are not the same, from the age of three or four – this is called the theory of mind.

The ability to pretend to be someone else in our mind's theatre also means we can imagine a different sort of self, the self that we would ideally like to be. Our own imagination may encourage us to practice spreading our wings a little and seeing how it goes.

The sense of self and the sense that "I" am in control of my actions is important to mental health. Elderly people in an old people's home fare better when they are given control over some aspect of their life, such as being expected to take care of themselves in some way or given a plant to take care of. Feeling that we are not in control is one way that life can lose its sense of meaning, unless we have willingly given up control for a good reason.

One aspect of ageing, of moving into the Crone years, is that this may happen whether we want it to or not. We are more likely to get into situations where we feel we are not in control and that we are not the author of our lives. If there is truly nothing we can do about it, it is best to decide to accept the situation and thus have a sense that we have willingly given up control over that particular aspect. Whatever happens, we always have the option of deciding how to feel about it – or what feelings to work towards.

Initially, however, any change will potentially make us feel that our sense of self has experienced a discontinuity, and that we are not in control, particularly if it is abrupt. The change may be physical, or in our circumstances. The people around us may act differently towards us, again threatening our sense of self. There may be a feeling that this particular situation has no meaning, that we can't get a handle on it, resulting in a sense of confusion or hopelessness.

If we find new ways of restoring our sense of meaning and the sense of authorship, this in turn may give birth to a new sense of who we are. And the archetypal way in which humans find meaning in their experiences is through the use of story.

Thus story is essential for humans in organising our experience and finding meaning and a coherent sense of self. However, many different stories might be told from the same set of events. Sometimes the stories we tell ourselves about our past are not helpful (we have concluded we are stupid or unloveable, or that the world is a threatening place we should hide from, or that we need to be aggressive because others are hostile etc). Equally, stories about the present may not be helpful. Since we construct our own stories, we can also rewrite them, by reinterpreting the raw data.

Narrative therapy can help to enrich our personal stories. For example, suppose you are ill. This may cause you so much distress, mental as well as physical, and be so hard to accept and adjust to, that it becomes your whole story. Who are you? An ill person, who cannot go on holiday any more and who must watch the garden become overgrown and who misses out on social occasions.

A narrative therapist would concentrate on enlarging this narrow version of your story, to take account of all the myriad things you have done and been in the past and your hopes and plans for the future, your loves and losses and rich interconnectedness of relationships, and who you are now – your interests and hobbies, your thoughts and dreams, your unique blend of traits and virtues and vices and habits and abilities. Just one part of this tapestry would be the thread – even if of a glaring colour and unusual thickness – of your illness.

Solution-focussed therapy might concentrate more on all the times in the past you have coped with adversity, and examine in detail how you have coped and endured and overcome and survived in large matters and small, over and over again throughout your life, to be who you are today. From this you extrapolate to a future where once again you have come through to find new meaning and purpose and joy. Your story is more positive and hopeful.

Since the story is constantly developing, and we are the authors of our own stories, we can take whatever steps we like to direct the story in a more helpful direction, so long as we remain true to ourselves – our sense of self must be reasonably coherent and continual. One way is to find inspiration from myth and legend and fairy tale, as well as other metaphors for life situations, and which seem relevant to our own life. They are a store of human wisdom, and we can borrow from that store for our own purposes.

Since we are all human, we all have a capability to be wise, courageous and kind, as well as selfish, weak and bitter. We will not be betraying our basic sense of self if we look to stories which help bring out the best of our qualities. When we read a story about a hero, we may accept that we are not brave enough to tackle a real-life dragon, but we can recognise that yes, we too have been in a difficult situation and faced up to it. We too have been scared and coped with it. We are all heroes in our own story.

CHAPTER 6: The Evolution of the Wise Woman: the evolution of ageing

In 1992 the Omega Institute for Holistic Studies in the United States organized a conference entitled "Conscious Aging" (US spelling). They invited speakers from the consciousness movement and leading figures from the study of ageing, and over 1500 people attended. Stephen Rechtschaffen, the Director of the Omega Institute, described conscious ageing as "a new way of looking at and experiencing ageing that moves beyond our cultural obsession with youth towards a respect and need for the wisdom of age."

The new approach to ageing was about striving for awareness as we grow old, as part of our spiritual journey. We are to recognise and accept the ageing process as a natural part of the life cycle, and not as something to be feared. It is also about changing the social attitude to ageing, seeing it as a time when the wisdom of the elders can be put at the service of society, for example by mentoring the young. Two of the most prominent members of the movement are Ram Dass, author of *Still Here: Embracing Aging, Changing and Dying*, and Rabbi Zalman Schachter-Shalomi, founder of the Spiritual Eldering Institute in the United States.

If we are willing to look closely at the matter of ageing and dying, we find it takes us on a fruitful journey towards the second major archetype of this book, the Wise Woman. Why is it exactly that we grow old and die? More specifically, why do women go through the menopause? Why does our fertility shut down so abruptly – and why do we live on for so many years afterwards? Is there any purpose and meaning in all this?

Answering these questions leads us to the conclusion that our bodily development shapes us to be in our later years holders of wisdom, knowledge and nurturing for the rest of the tribe. It seems that women at least do age in order to become sages. But we also have to ask: why is the Wise Woman not respected today? If our very biology makes us Wise Women, why is this archetype now hidden from us?

The theory of evolution explains that those best adapted to their environment are better at producing young and thus passing on their characteristics. Charles Darwin first used the term "natural selection" to describe how traits that allow some individuals to produce more offspring will spread through the population, while those traits that cause an individual to perform badly will be lost. An individual might for example be faster and better at catching prey or avoiding predators, or might be more resistant to disease, or more attractive to potential mates. Others with less desirable traits may die off sooner or fail for other reasons to produce as many viable young. Eventually the population will consist only of the descendants of the better adapted individuals.

We now know that the differences between individuals - when not caused by the environment itself, for example being hurt in an accident - are caused by having different genes. (Genes consist of a stretch of DNA which is found in the chromosomes of our cell nuclei.) Thus genes that give an advantage to its owner will tend to spread through the population. We might ask: how could ageing and death possibly have evolved through natural selection? Surely any organism that did not age and die would produce many more offspring, and thus become the common type?

As we look around us, we see that each species has a natural life span and that these differ enormously. We expect to reach our threescore and ten years, perhaps even 80 or 90; some reach 100 but we would be surprised to find someone of 110. Nobody makes it to 130. Sequoia trees on the other hand live for 2,000 years or more, and don't show signs of senescence (changes due to ageing). A cat might last 20 years and an adult moth may live for a few days.

When we examine the fate of animals in the wild, we find that very few, if any, meet death through 'old age'. An animal may meet death through accident or disease, and as soon as it becomes weak it is likely to fall victim to predation. Our own cells and bodily organs are also subject to wear and tear, disease and accidents, and eventually weaken and die, making the whole body weaker (old age, senescence) and thus more likely to die.

One early observation was that larger animals tend to live the longest. Whales, for example, may live long past 100 years of age; guinea pigs maybe six years; large clams can live to 40 plus. Also, animals that live in colder places like the Antarctica can live longer – gobies and blennies may live three or four times longer than related fish in warmer climes, and reach 40 years or more.

From this evolved the "rate of living" theory. This suggested that ageing was related to the pace of life – where there was a lower body temperature, a lower metabolic rate, a lower growth rate, or less activity, the animal would live longer. It was suggested that we all had about the same number of heart beats in a lifetime before we wore out – but a mouse's heart would beat very quickly for a short time, while ours would beat much more slowly for a long time since larger animals have a lower metabolic rate.

However, there were too many exceptions to this rule. For example a humming bird has a very high metabolic rate and is very small, but can live up to 14 years. So this theory can't explain everything.

Another way of looking at life span considers the payoff between concentrating on body quality or on reproduction. Since the body is bound to age, and likely to die eventually anyway by some chance event, there are limits to how useful it is to pour resources into trying to stay alive. Once you're dead, that's it in evolutionary terms, unless you have managed to leave offspring behind. Any new gene that enabled you to live for 500 years would be lost for ever if you haven't managed to pass on that gene to your descendants.

Living longer means that the damage that is constantly occurring within our cells and tissues has to be more efficiently monitored and repaired and rebuilt. We take for granted the miracle that says if we cut ourselves, then the blood clots, so we don't lose too much; and that our immune system fights off the bacteria that get into the wound; and that new skin appears to seal off the body and protect us from further harm. Or that if we break a leg, it heals. But mostly we don't see what is going on with our various protection and repair mechanisms, and don't realise how costly it is in terms of bodily resources. Eventually however, the protection and repair mechanisms themselves age and become less efficient.

Given that resources are generally limited, there is a tension between putting energy into staying alive – improving the quality of your bodily systems, finding new ways to fight off deterioration and disease – and putting energy into reproducing. You might get ahead if you concentrate on copying your genes into the next generation as soon as possible. On the other hand, if you are too careless about your body quality you may die before you get a chance to reproduce, or you might lose your chance of mating to a more attractive rival.

Thus it's a question of a delicate balance between life span and reproduction, and clearly different species have responded to the problem in different ways.

Animals can be roughly divided into two groups: r strategists and K-strategists. The r-strategists go for quantity. They have relatively short life spans, produce large numbers of offspring at a time and perhaps reproduce several times a year. The offspring tend to be very vulnerable and are left to look after themselves; probably many die but so many are produced that there are likely to be descendants. Think of the millions of eggs produced by fish, or the many litters that can result in one year from a starting point of one pair of mice.

The K-strategists use their resources in the pursuit of quality. They have longer life spans, larger bodies and fewer offspring, but ones that are better equipped for survival. They give their offspring a good start in life. They are less vulnerable to the environment and although their offspring are few, they are far more likely to survive. Humans are the prime example of the K strategy, and we have a relatively long life span.

It's not that one strategy is better than the other – both rats and humans have spread all over the world. It's a question of which path has evolved for a particular species. The human brain gives us a good chance of living longer than one might expect, given our size, so we can take the chance of a long childhood before becoming sexually mature. That long period of maturation allows the brain to reach a state of development that will put us well ahead in the survival stakes, giving plenty of time both to produce children and give them a good chance of reproducing themselves.

Often species that live relatively long lives tend to be well protected from predators. Once past the vulnerable chick stage birds can fly out of trouble. Bats live on average three times as long as similar sized mammals that don't fly. Clams and tortoises have tough shells. Large animals such as elephants are protected from predation by their size, whereas mice are at high risk of predation. Living together as social animals increases the chances of a longer life for the group – we are an example of a social animal.

So just what is going on as our bodies age? One theory is the 'genetic dustbin' theory of old age. The geneticist JBS Haldane realised that if a gene has a bad effect that only manifests itself later in life (eg Huntingdon's disease in humans), the chances of it being removed through the process of natural selection are slight. The possessor of the gene may well be already be dead, due to accidents, predation, disease or starvation before it is switched on. Most if not all of the production of the next generation would probably already be over and thus the gene would already be passed on.

Peter Medawar, an immunologist, applied this insight to the theory of ageing. He realised that the body would gradually become less and less efficient, because the power of natural selection to remove particular genes was lessened as their deleterious effect was turned on at later and still later stages in our life. A small decrease in bodily efficiency might have a great effect on relative reproductive success if it appeared in youth. If it appeared in middle age, the disadvantage would need to be much greater to have much impact in evolutionary terms, since offspring would already have been produced and the defective gene passed on. Still later a bad effect would have no impact at all.

For example, a version of a gene that causes us heart defects is likely to selected against in humans (in our original hunter-gatherer state) if its effects appear in early adulthood, so we can generally expect to have strong hearts in our youth. Deleterious effects appearing in our forties may well still affect our success in raising the offspring we have already produced, so again we can expect evolution to have provided us with reasonably strong hearts, though maybe not as vigorous as before since we have already produced children and passed on the gene. Deleterious effects in our eighties will make no difference to our reproductive success, the relevant genes having already been passed on to our descendants, and natural selection will not protect us.

Multiply these effects for all our bodily systems and you have the effects of ageing and senescence. Since mutations are occurring all the time, and most mutations are harmful, it is not surprising that so many bad effects have accumulated to let us down as we age. Genes often have multiple effects, and it may well be common that a gene with deleterious effects in old age, has actually been selected for because of its advantageous effects at a younger stage. Thus the term 'genetic dustbin' effect: because of the way natural selection works, all the bad effects of genes accumulate as a kind of rubbish at the later end of life.

What is actually going on at the basic level of the cell as we age? There are many causes of damage to cells and tissue but one major source is that done by free radicals.

All the complex processes of life require energy. We consume energy in the form of food, but this then has to be converted into a form which our cells can use in its various chemical processes. To put it very simply, we burn food with oxygen to produce energy – glucose plus oxygen produce carbon dioxide and water plus energy - but oxygen can be destructive. Sometimes some of the oxygen molecules escape and turn into free oxygen radicals, or oxidants, which are very reactive. As they damage other molecules, these in turn also become oxidants and cause more damage, and so on. Other processes in the cells produce other kinds of reactive radicals, such as hydrogen peroxide.

Various parts of the cell can become damaged by these radicals, but one of the most important parts is the DNA itself. Damage here causes mutations that will then produce defective enzymes, which will in turn cause more problems to the cell. No animal can avoid the production of radicals, but they can produce their own antioxidants to neutralise the radicals, and also invest in a complex system of repair. For example, damage in the DNA can be monitored and faulty bits snipped out and replaced. The more energy put into this process, the longer the animal is likely to live, barring accidents, predation etc.

Tom Kirkwood, Professor of Gerontology at Newcastle University, has put forward the 'disposable soma' theory. Ultimately the body (soma) is disposable, since it is the germline (reproduction) that is important. Ageing results from the accumulation of damage in the cells and tissues, due to free radicals and many other processes. We have evolved a particular level of ongoing maintenance and repair of the body, which is limited due to the competing need to invest in reproduction. Thus the main genes that determine how long we are likely to live and how quickly we age, are those concerned with maintenance and repair. The specific amount of damage done to our body, and thus the speed at which we age, is due to chance and environment at the level of the individual, but within the limits of our species' general lifespan.

Thus our genetic make up as a species, and to some extent as individuals, will help determine how long we are likely to live, but we can in theory extend our lives by reducing our exposure to damage (eg by environmental mutagens such as radiation and damaging chemicals) and improving our cell maintenance and repair systems.

Why does death exist at all? Can we envisage a planet on which death is unknown? Could not things have been arranged differently? When we look into these questions, we find that we would not exist were it not for death. Life arises out of death: out of the death of stars, from natural disasters such as volcanic eruptions and floods, and through the processes of evolution.

Scientists have discovered many of the laws of our universe over the last two or three hundred years, going ever deeper over the last century. They have found that the universe is finely balanced to make it inhabitable for life. A few more or less particles at the time of the Big Bang, slightly different figures for the various mathematical constants of the universe's physical laws, and we wouldn't be here. No life (as we know it) could be here or anywhere.

Some people take this as meaning that the universe was divinely constructed for the purpose of producing life and, in particular, us. Some mathematicians see it as meaning that there must be many universes, perhaps an infinity of them, and we naturally find ourselves on the one that allows us to be here.

As the universe ages, there is constant evolution or unfolding of the consequences of the Big Bang. This involves birth and death and rebirth, so that for example suns are born and make new elements in their nuclear furnace, then they die and explode and the clouds of their remains become incorporated in new suns or are carried through space. Thus the planets when born contain the elements from star dust that can make up life. Without the death of suns we wouldn't be here. On the other hand, our own sun will eventually die and if we haven't moved on by then we will all die too.

Equally evolution of life, by natural selection, means death. Life could perhaps have emerged in the first place without death, in the form of blue green algae and bacteria which was all there was for millions of years, but for the runaway progression of ever new forms that characterises the last 600 million years on earth, there had to be natural selection and that means death. If nothing dies there is no selection of particular types and therefore very little change. Quite soon all available ecological niches would be taken and that would be it. Certainly an absence of death means no animals, for they live by eating plants or each other.

As life forms became complicated death became ever more likely. The law of entropy states that matter seeks the lowest form of energy and everything tends towards the most simple state. It takes energy to keep complication going, energy that in our case comes ultimately from the sun, so that plants grow, and animals can eat them. The breakdown of the complicated state of life is likely to happen by chance at any time: accidents will happen. Even if you were immune to the internal processes of ageing you would die eventually, simply by one big accident or by the gradual breakdown of your body by many little accidents.

Since we have offspring we don't need to pour all our energy into the hopeless task of trying to keep our bodies alive for ever. The genes that code for our bodies get passed on to new and vigorous young bodies. And we die: as Tom Kirkwood put it, we have a disposable soma. Because some organisms are better fitted to their environment, survive better, and produce larger numbers of healthy offspring, the genes that coded for that fitness get passed on in larger quantities than other versions, and spread through the population. Eventually the less useful versions may get lost altogether. Over the countless millennia life evolves until oak trees, petunias, butterflies and humans appear. Had there been no death there would have been no evolution, and we would not be here.

The average length of human life in the developed world is steadily increasing, but there is a limit to how far that can be extended. We still age and in the end we shall all die. This is the price we pay for living in the type of universe that we do live in. If there was no death, we would not be here in the first place to fear it. Even taking a comparatively tiny segment of life's history on earth, imagine the changes through time from the australopithecines, the ape which first walked upright, to *Homo habilis*, to *Homo erectus*, to archaic *Homo sapiens* and the Neanderthals to the first modern humans, to you. Had there been no death, there would have been no selection and no evolution, and you would not exist. You cannot, in this world, do away with the fact of death without doing away with the fact of your life.

When we look at the things that kill us, apart from accidents and old age, we still find that life goes hand in hand with death. Diseases and parasites are forms of life – bacteria, viruses, protozoa, worms, insects – they all come from the tree of life that also produced ourselves. Without starting off with the simpler forms such as bacteria which may now prey on us, there wouldn't have been the evolution of more complex forms. Without insects (some of which are a nuisance or even deadly), there would be no pollination for many species and we would go hungry. We live in an intricate web of life and if you disturb one bit, another bit trembles.

Even the natural disasters such as tsunamis and volcanic eruptions, earthquakes and floods are, as a whole, part of life. The continual turnover of new materials into and from the interior of the earth provides new minerals and fertile soils, although the side effect of plate tectonics (movements of the earth) is that sometimes it kills us. The heat of volcanic activity in early earth history may have helped the initial evolution of life on earth. The many different ecosystems caused by the restlessness of the earth's surface has helped lead to the complexity of the ecosystem which produced humans, among others. The major climate changes that led to the retreat of the forests in the Miocene meant that the first steps from ape to human took place as some apes came down from the trees. Floods refertilise the soil, and helped bring about the first civilisations such as Sumer around the Euphrates and Egypt around the Nile.

The way the universe is constructed means that we wouldn't be here, and couldn't remain here, if the universe were any different to how it is. That is inevitable of course, since we are adapted by evolution to fit in to the universe in which we find ourselves. Whether the universe could have been constructed in any other way is as much a religious question as a scientific one.

CHAPTER 7: Evolution of the Wise Woman: the evolution of the menopause

There is a major problem still to be explained. The driving force in evolution is always reproduction, the passing on of our genes, the unwitting competition to leave more offspring than our rivals. So how do we explain the menopause, the abrupt cessation of reproductive ability? How do we explain that women then go on to live for perhaps thirty or forty years, even though they can no longer produce babies? Shouldn't natural selection choose women who can have babies? Alternatively, why don't we die sooner?

Women's monthly cycles cease because we stop producing eggs. All the eggs we will ever have, about 7 million, are made while we are still in the foetal stage, and they immediately start dying. By the time we are born, our ovaries may contain about 1 to 2 million eggs. By puberty there are about 250,000 left, and we start releasing them monthly for potential fertilisation. Each month one develops and is shed to move down into the womb, while many more die. At the age of about 35 the rate of loss speeds up, and most are gone by menopause.

Since oestrogen is produced by the cells that surround the developing eggs, when few are left the level of oestrogen falls so that the monthly cycle (which it controls) can no longer be sustained. Thus egg development stops, levels of oestrogen and progesterone fall even more sharply, and fertility is lost. This does not happen with the female of many other animal species since they do not stop making eggs in the foetal stage and therefore do not run out of them.

We have seen how important reproduction is to evolution, and how natural selection favours those genes that lead to the successful production of offspring, to the extent that we can talk of the 'disposable soma' – the body only counts as an efficient vehicle for passing on its genes. How, then, could a system evolve such that female humans lose their ability to reproduce while still potentially active and healthy, and continue to live on for perhaps another thirty or forty years? It is not uncommon for a woman to live one third of her life in a non-reproductive state. How could this have evolved in the first place? The menopause is not found in chimpanzees for example, which suggests it was not present in our common ancestor.

It seems unlikely that this is one aspect of the 'genetic dustbin'. The timing of the menopause suggests that it is genetically programmed rather than part of the general deterioration of ageing. Most women experience their menopause at around 50 years of age, with most variation falling within about five years either side. This is far more precise than the human life span.

If the timing is genetically programmed, that suggests that it has been selected for by natural selection – that it has evolved because somehow undergoing menopause confers an evolutionary advantage. And yet the menopause means that a woman produces fewer children than she might otherwise have done, which means fewer opportunities to pass on her genes. If a woman had genes that enabled her to continue to be fertile throughout her life (as men can be), surely those genes would be passed on more often, so that menopausal genes could never become established.

Either the theory of evolution by natural selection is wrong, or there must be some other explanation. One suggestion is that menopause is simply an artefact of modern living. In pre-technological societies, the argument goes, hardly any women would live long enough to experience the menopause, since practically all of them would be dead by 40. Any talk of timing of the menopause is irrelevant because women would not reach that time anyway. Incidentally, this argument would mean that the female body is not adapted to undergo the loss of oestrogen and progesterone, and the case for HRT (hormone replacement therapy) is strengthened.

This theory is based on the study of long dead human bones which suggest a particular age structure of ancient peoples. New evidence suggests a different picture; *average* life expectancy at birth may well have been low, but once past the hazards of infancy where disease strikes hardest, life expectancy would then be much higher. The Psalms in the Bible were composed in a pre-technological society, but Psalm 90 famously states: "The length of our days is seventy years – or eighty, if we have the strength." The argument also fails to explain why the reproductive system in a woman should age so much faster than her other bodily systems, so that she can now expect to live in reasonably good health for some time after the menopause; and so much faster than a man's reproductive system, although on average she outlives him.

Another explanation of the menopause is that evolution through natural selection requires not only that you reproduce, but that at least some of the offspring survive to reproduce in their turn. If an ageing woman gives birth but then either dies of old age or is too frail to properly care for the child, then there is little to be gained; on the other hand, if she ceases to undergo the exhausting business of childbirth, she can make sure that her last children survive to maturity.

It seems that menopause can be observed in some other species, for example in whales. As the years go by, it seems to be less uncommon than first thought. Craig Packer, working with lions and baboons, has shown that (barring death from causes other than old age) females can expect to live long enough after the birth of their last infant to successfully rear them. Orphaned infants are likely to die.

Packer estimates that equivalent timing in human females would explain their survival to about 58 to 65 years, assuming that human infants needed their mothers to the age of 10 years to survive. (This seems a little unreasonable, since few women become mothers at 55.) At any rate, most women would have a natural life span of only 60 years, which is clearly not the case. Today life expectancy after the menopause is not ten years but as much as thirty or more. Thus our menopause is something quite different.

So how else can genes for *not* having children get passed on? The problem can be approached by looking at other situations where a gene seems unlikely to get passed on but has indeed possessed an evolutionary advantage. There are two common situations: reciprocal altruism and kin selection.

We can see many social species where one member of the group will help another, at some cost to themselves (known as altruism). Food sharing is an obvious example: instead of wolfing it all down, or hiding some for later, one member of a group may offer some to another. More complicated altruism takes place when, for example, one chimpanzee will help another to overcome a rival and become the alpha male, risking injury in the process although they will not become the alpha themselves. Surely giving up food, or risking injury, means the possessor of the altruistic gene is less likely to produce and successfully rear offspring, so the gene should be lost?

Reciprocal altruism explains that a later pay off gives the altruistic animal an evolutionary advantage. Although it costs some of that animal's resources to help another, if he or she is likely to see the favour returned then both will be better off and both will pass on the genes promoting such behaviour. The one sharing food now may be glad to have the favour returned when times are hard. Chimpanzees helping an ally to a higher position, with its greater access to mates and resources, can expect to be granted extra privileges, just as a newly-crowned human king might bestow lands and titles on a knight who helped him to victory over a rival. Both chimpanzees will have increased their chances of reproducing and the genes responsible are more likely to be passed on than non-altruistic genes.

This shows that evolution is not necessarily about "nature red in tooth and claw". Competition is often stressed as the be-all and end-all of natural selection, but evolutionary success is about good adaptation to the environment. Social groups with an emphasis on cooperation can be enormously successful: think of humans, but also dogs, ants, termites.

We have reached the point where most of us would find it impossible to survive on our own. As well as depending on joint efforts to produce food and warmth and shelter, few of us could stand the loneliness. Once the most important part of your environment is other people, being without any genes for reciprocal altruism could make you friendless, disliked and at risk of eviction from the group (or even murder). Pure selfishness would then be a liability in evolutionary terms.

Although reciprocal altruism is of immense significance to human evolution, it does not seem to offer any explanation for the menopause. We will turn to kin selection. This refers to the fact that it is the gene that is being selected in the course of evolution, not the individual (who dies and disappears for good), and that if one member of a species has a copy of a certain gene, it's likely that its relatives (kin) have a copy also. Your children, for example, have half of your genes and half of your partner's genes.

Thus a gene that causes you to give up resources or even your life for your family has a good chance of being passed on. If your sacrifice makes them more likely to survive than another family without the gene, your genes are likely to prevail in coming generations.

It obviously makes evolutionary sense for a woman to give up her life for her children under certain circumstances. While there is the risk that they won't grow to maturity without her, if they will die without her sacrifice then her line is finished anyway – she has failed to successfully reproduce. What the theory of kin selection shows is that this kind of altruism goes beyond one's own offspring, and can apply to members of the wider family. If you risk your life to save your niece, the gene that caused you to do so is quite likely to be passed on through her, since you and she are related. If you die to save a whole extended family or tribe, the altruistic gene will almost certainly be passed on, since it is likely that at least some of the tribe will be carrying that gene.

Kin selection is found in many species, not just humans. It even explains the odd case of worker bees, who give up all chance of reproduction to tend the offspring of the queen. How that could evolve through natural selection was a puzzle, until it was noted that by a particular quirk of bee reproduction, worker bees are genetically closer to the offspring of their sister – the queen bee – than normal and actually have a greater chance of passing on a particular gene by raising their sister's children than they do from raising their own.

Let us see how kin selection might apply to the menopause. Might it be that a woman can promote the chances of survival of her kin by giving up the chance of reproducing past middle age? And would the effect be so marked that it would compensate for this loss of fertility?

If a woman continued to produce healthy babies and raise them to maturity as she entered old age, kin selection would not make sense. The surest way of passing on your genes would be to produce your own children. However, we need to consider the fact that a woman's body is already showing signs of ageing by midlife, and that pregnancy, childbirth and lactation are exhausting and often risky processes. Men on the other hand can impregnate women with very little physical cost to themselves, and can continue to do so into old age.

Due to the accumulation of internal and external environmental 'insults' (damage of various kinds) that inevitably happen, a woman of 50 plus – particularly in an ancient society – is likely to be less fit, healthy and energetic, less likely to produce healthy eggs and less likely to survive childbirth. If she dies in childbirth her younger children may then be at risk and fail to make it to maturity. It may make evolutionary sense to give up having babies herself, and concentrate on the next generation – which will be carrying her genes.

Thus there has developed the "grandmother hypothesis", which now has many proponents, one of the most prominent being Kristin Hawkes of the University of Utah in Salt Lake City. She suggests that the key point is not early loss of fertility, but the unusually long life after the menopause, which has been selected for in human females. She sees the reason for this being down to mother-child food sharing, particularly of hard-to-gather food such as deeply buried roots and tubers which would have been important when we were hunter-gatherers.

Once past the menopause, a woman had no young children of her own to forage for, so she foraged for her daughters' and perhaps nieces' children. Studies of the strength of modern foraging women has shown no marked decline till well into the sixties. By increasing the chances of survival of her grandchildren, the chances of her own genes being passed on (including the genes for menopause) are also increased.

In fact a grandmother may well be able to enhance her grandchildren's survival in a number of ways. If you have any experience of small children, you will be aware of their curiosity and ceaseless activity – exploring, touching, fiddling with things. You have to keep an eye on them all the time to make sure they don't fall to their deaths, poke a finger into a light socket, eat something poisonous, wander off and get lost, or drown in a foot of water. Whereas in early days there would be no light sockets, there would be a greater danger in getting lost, or being seized by a hyena. Children's curiosity reflects our human intelligence, but it does make close supervision for a number of years an essential part of parenting.

With a grandmother or great aunt to help supervise the small children, their chances of survival must have been increased, quite apart from the extra food that could be gathered for them. The extra attention would also enhance the training and socialisation of the children, as skills were taught and questions answered or even encouraged. The grandmother role would be particularly important as the mother gave birth to a further child, and was much less available to her existing children. Children with grandmothers could well have been more fit than those without. Of course many women would have died by chance before menopause or shortly afterwards, but enough must have survived to help spread the genes for menopause through the population.

To the role of grandmother I would add the role of wise woman. There is a limit to how many children that one woman can help feed or supervise at a time. However, a major characteristic of humans is our culture. We do not depend on instinct for our survival, but on learning and the passing on of skills and knowledge. This is what has enabled humans to spread all over the earth, even though we were originally tropical animals. Our culture is freed from the constraints of biological evolution, which is very slow compared to the human life span, and enables advances in adaptation to progress incredibly rapidly. The human species has only been around for a blink of an eye in evolutionary timescales, but already we can visit the moon.

Today our knowledge is written down. In massive libraries and on the internet, there is vastly more knowledge available to the human race than any one human could absorb. For most of human history, that was not the case – knowledge was carried in the brain and passed on directly from human to human. That knowledge came in many different forms, but most important would be knowledge of how to physically survive.

This would include knowledge of different foods, where to find them, when they were in season, how to gather them or hunt them, what should be avoided because it is poisonous, what circling vultures might mean about meat to scavenge, what doesn't taste very nice but can be fallen back on when times are hard, which animals are dangerous and how to avoid them, which herbs are useful in childbirth or illness or injury, different sources of water, how to make fire, how to make tools, how to construct shelters or where to find caves, in different places according to where you are gathering or following food at the time; how to make coverings that will stay on you, bags to carry what you have caught or gathered – the list must be enormous.

Hunter-gatherer societies can be very sophisticated, for they must be in order to survive as well as they do, and that depends on the ability to learn knowledge and skills, to make sense of and classify knowledge, to prioritise what needs to be taught and to understand how it all fits together.

Although we can't know much about our ancestor's daily lives, it seems likely from contemporary hunter-gatherer societies that there was some gender division of labour, if only because it is women that bear the children and breastfeed them for several years. Men were probably more likely to be involved in hunting and associated tool making, while women were more likely to be involved with gathering and processing of food while minding the children. Beyond that it is hard to say. It does seem likely that older women would have absorbed a life time of knowledge and skills that were priceless in terms of their value to the group. In a sense they would be grandmothering not just their own grandchildren, but their whole tribe.

There must have been times when an old woman's knowledge of where to find alternative sources of food when an expected source failed, or any of the list suggested above, made the difference between life and death for the whole group or tribe. Most of the tribe would be related to her quite closely, and therefore saving the tribe would be saving the genes for menopause. A longer life span after menopause and freedom from the hazards of childbearing as her body became less able to cope, a reduction in essential duties enabling her to exercise her knowledge and skills and gain further knowledge and wisdom, would all make an older woman not just a useful grandmother, but a valuable tribal resource – the Wise Woman.

Humans are omnivorous, and we used our large brain partly to amass a huge range of foodstuffs. As well as meat (catching small animals as well as scavenging and hunting large ones), eggs, shellfish and so on we originally ate many more different types of fruits, vegetables, roots, fungus, seeds and nuts than we do today. During the excavation of one of the first settlements in the Middle East, shortly after the last Ice Age, about 150 different plant species were found to have been gathered and brought back to the village. Because of the lottery as to which pieces of evidence survive and which are lost, archaeologists estimated that this probably meant more than 250 species had actually been gathered. Had the people involved been true wandering hunter-gatherers, the total would perhaps have been much more.

It is hard for us today, especially in the prosperous West, to imagine the importance of that sort of knowledge. If we want food, we go to the local supermarket. For water, we turn on the tap. The knowledge we use is for far less basic purposes and therefore less of a life and death issue. The knowledge accrued over the decades by a woman hunter-gather would have been immense, and of immense importance. For such a woman to survive into old age and to continue to share her knowledge with her little band could easily make all the difference between survival and the whole tribe being wiped out in a bad year.

The tremendous evolutionary advantage to be gained from such knowledge and skills remaining at the disposal of the tribe, in the days before writing, must surely have made the menopause a prime target for natural selection. Even if only a minority of the women carrying menopausal genes survived to old age, their descendents could flourish. And the knowledge they carried in their heads, and the extra care they gave to the children of the tribe, could well have repaid any physical care and resources given to an aged and infirm woman by the rest of the tribe.

We take for granted the archetype of the wise older man, the grandfather – Merlin or Gandalf. We accept that the knowledge and skills of older men were useful to society, and that often the male Elders had great authority in their tribe or clan. We have forgotten the time when women Elders were also important, but there are many examples in modern or recent hunter-gatherer societies (eg see Leacock, 1981). We shall examine in later chapters how long women were able to hold power and influence.

No-one can yet suggest with any confidence when exactly the menopause as we know it today first evolved. Kristin Hawkes and her colleagues suggest three important transitional times when it might have happened – the appearance of our ancestor, *Homo erectus*, about 2 million years ago, who was successful enough to spread out of Africa; the first appearance of archaic *Homo sapiens* (called *Homo heidelbergensis* or *Homo sapiens heidelbergensis*, depending on the scientist) in Africa about 600,000 years ago; or with the dispersal of fully modern humans out of Africa and across the continents about 90-50,000 years ago.

Four villages in rural Gambia have been studied by the Human Evolutionary Ecology Group, in association with Tom Kirkwood and colleagues at the University of Newcastle. They found that older women do help in raising children and that having a maternal grandmother significantly improves the chances of a child's survival. A mathematical model built to look at the trade-off between continuing to produce children with all the costs and risks of childbirth at a later age, or concentrating on the offspring of the children you have already produced, did predict the evolution of menopause.

It is interesting that this model, concentrating only on the care that a grandmother gives her grandchildren, predicts a menopause starting at a later age than actually found. In other words, the model finds enough benefit to give a selective advantage to a woman starting menopause, but not until she is older and has had more years producing her own children than actually happens. This may well be because the "wise woman" effect was not considered by these researchers.

The extra role of the postmenopausal woman, the tribal Elder, helping not just her daughters to raise their children, but helping the whole tribe in a variety of ways through her experience and wisdom, is far more difficult to measure. However this could well have been of such importance to the survival of the whole tribe that it compensated for earlier loss of fertility. By having a shorter time period in which to give birth to their own children, women contributed to the survival of all the genes of her tribe, including the genes for menopause. Thus such a tribe was more likely to survive than one where the women did not undergo the menopause, or tended to go through it later.

The grandmother effect is usually about maternal grandmothers. This suggests that in early human hunter-gatherer societies daughters were more likely to stay close to their mother (matrilocality), while sons would move to be with their "mother-in-law". Thus matriliny – measuring descent through the female – would be far more logical than patriliny, until civilization changed the rules. And as we shall see, matriliny and a higher status for women tend to go together.

Further research looking at infant survival in Germany (16th - 19th century) and Finland and Canada (18th-19th century) also showed a maternal grandmother effect. The data showed that women had two extra grandchildren for each extra decade they lived after 50. Women were more likely to have children at a younger age if their own mother was still alive, and the grandchildren were more likely to survive. Children who reached the age of two were more likely to reach the age of five if they had a living grandmother. The younger the grandmother, the more likely the grandchild was to survive. The researchers pointed out that mortality rates in grandmothers accelerated at the same time that their own daughters were ceasing to reproduce and about to become grandmothers themselves.

Grandmothering continues to be important in the West today, although of course not in the same way as in hunter-gatherer societies. According to Age Concern, in six out of ten families in the UK today, the main source of substitute care for the children is grandparents, usually grandmothers. This is in spite of the change in emphasis from the extended to the nuclear family.

The biggest objection to the grandmother hypothesis has been the idea that our ancestors simply didn't live long enough to experience the menopause. This objection carries much less weight today. People became sceptical when it was found that many contemporary pretechnological societies live much longer than the original evidence suggested they should. For example, 40% of the women of the !Kung people of Africa reach their menopause, generally in their mid-forties. One study found that in 14 different ethnographic censuses, almost one third of adults lived beyond their mid forties. Among the Ache, 39% lived to over 45, among the Hadza, 40.7% reached 45 or older.

An average life expectancy of, say, 40 years in earlier societies does not mean that few people survived past 40 years. Once past the hazards of infant diseases, and then the hazards of childbearing, the survivors might then live to a ripe old age. It is not so easy to tell the age at death from the remaining bones, and it has been shown that data has been misinterpreted. In particular, skeletons survive differently according to their age at death. Records from a mission cemetery in California less than 200 years old showed that 53% of the adults buried at the cemetery were over 45 years, but only 7% of the recovered adult skeletons were over 45. The rest had obviously rotted away more quickly than younger, stronger bones. There may well be even greater discrepancies if the bones are thousands of years old – and perhaps suffered the effects of osteoporosis at the time of death.

Thus it seems likely that a significant number of our foremothers survived as much as 30 years after they had finished producing their own children. Barring accidents, they remained active members of the group, playing essential roles in the life of the tribe. So important was their role, that evolution had freed them from the hazards of childbearing to take up a new role – that of grandmother and Wise Woman.

CHAPTER 8: Decline of the Wise Woman

Most of us would probably agree that although there are still some opportunities for the role of the Wise Woman, that role is very limited today. We are still often wanted as grandmothers (though increasingly, not even for that role), but wisdom is not now associated with old women. On the contrary, the social stereotype is more likely to be of some silly, narrow minded "old bag". "Old wives' tales" are supposed to be untrue. And as the old saying goes: now I have all the answers, nobody asks me the questions.

Why has the role of the Wise Woman declined? What happened to take away the respect of the tribe? When did it all change, and how?

The first writing was invented about 3,000 BC. Before then – prehistory – we can only learn about our ancestors from the material they left behind them: the remains of their skeletons, burials, shelter, tools, pottery, figurines, food remains, jewelry, art. Very little of what ever existed remains available to be dug up or found by archaeologists, and certain items such as woodwork and clothing are likely to decay and never be found.

Exact dates are not known, but it is thought that modern humans – *Homo sapiens sapiens* - first arose in Africa about 120,000 to 100,000 years ago; moved out to the Middle East by 90,000 years ago; to the Indian sub-continent and beyond by 70,000 years ago; and to Europe about 50,000 years ago. The Americas were not reached until about 20-15,000 years ago, and various islands later than that. Everywhere modern humans replaced the old, including our cousin, *Homo sapiens neanderthalensis*, who arose as a cold-adapted hominid during the Ice Ages. (Marked climate fluctuations began about 2 million years ago.)

We find evidence of a much richer culture with the rise of modern humans: particularly after the start of the Upper Paleolithic period (Old Stone Age), about 60-40,000 years ago. Evidently there had been a significant change in the way our minds worked, as well as an increase in brain size.

The innovations first seen about 60,000 years ago had spread all over the inhabited world by about 30,000 years ago. These included art, beautiful and highly sophisticated as seen in cave art; sculpture of stone and bone; a flowering of more effective strategies in hunting, with a wide variety of tools for the various jobs; diversification of tools, such as for sewing clothes and grinding plant material; constructions of, for example, dwellings made of mammoth bones; art being used to store information, such as tally marks (and the cave art could have been a teaching device, as well as of religious significance); the start of religious practice, such as elaborate burials with grave goods, and the possible depiction of shamans; and personal adornment, which also suggests recognition of different social roles.

Clearly by 30,000 BCE these were fully modern humans, just like ourselves. Though living a different way of life, these people were no less intelligent than us, and their way of life and social system may have been just as complicated. A burial at Sungir in Russia dating to 28,000 years ago contains the graves of an older man of perhaps 60 years and two adolescents, male and female. Each was adorned with strands of literally thousands of beads, plus bracelets, necklaces, decorated belts, arctic fox teeth, ivory pins, discs, pendants and lances. In spite of the harsh climate, these people had time and resources beyond that needed just to scrape a living.

Since all human females undergo the menopause, it must have evolved before or with the rise of the first fully modern humans. The grandmother effect may have been useful as far back as the Pliocene era (5-2 million years ago) as the climate cooled and much of the fruitful forests of our ape ancestors were replaced by grassy savannah. The Wise Woman effect, while always useful to an omnivore, would be particularly significant in the far more complex world of the modern human, in terms of technology, knowledge of natural history, and the intricacies of social relations.

As time went by, cultural innovations slowly increased. There is evidence of gathering places where hunter-gatherers from a wide area would live together for perhaps several weeks or longer, exchanging information and skills as well as sharing food and goods. But we remained hunter-gatherers through the millenia, until after the last Ice Age. The Last Glacial Maximum (LGM) occurred at about 20,000 years ago. The world population was small, and much of the world was either covered in ice sheets or in conditions of arctic tundra. Since conditions were also dry, drought was widespread, and low sea levels exposed often barren coastal plains. People did survive, even in harsh conditions, though none could penetrate the ice sheet.

Global warming began, with fits and starts, after 20,000 BCE and the climate became warmer and wetter. By 15,000 BCE the ice sheets were melting and conditions were good by 12,700 BCE. A hiccup then occurred (the Younger Dryas) from 10,800 BCE – 9,600 BCE, when the climate sharply deteriorated again to being cold and dry; after that, there was rapid warming again and the Ice Age was over. We are now in an inter-glacial period, and the earth will tip over into another ice age sooner or later due to fluctuations in the earth's orbit and other factors.

A major revolution in human history now occurred – the Neolithic revolution (New Stone Age). After millions of years of history as a hunter-gatherer, including tens of thousands of years as a fully modern human hunter-gatherer, for some reason we decided to settle down and become farmers.

Hunter-gatherers may have temporary settlements, but generally are on the move. This is partly because they gather or use up all the resources in one particular area and then move on to another; or they move to take advantage of seasonal variation – plant foods becoming available at different times, or animals migrating at certain seasons.

As the climate became warm and wet after the last Ice Age, the living became easier. Plants and animals suitable for food became abundant. Early communities of hunter-gatherers (called Natufian) are found from around 12,500 BC in the woodlands of the eastern Mediterranean, such as at Ain Mallaha near the Sea of Galilee. Steven Mithen describes how all the villages are found at the junction of thick woodland and forest steppe, providing food from two different habitats and probably near water. Gazelle bones are found in abundance, as well as deer, foxes, lizards, fish and game birds. The teeth of the gazelles, due to seasonal variations in their growth, show that whereas in earlier sites there was seasonal occupation, these people probably remained in their village all the year round. They were some of the first sedentary humans.

Many rat, mouse and sparrow bones are found, which also suggests a permanent village. Dogs had become tamed, perhaps used as guard dogs. Flint blades have been found with 'sickle-gloss', which indicates that they had been used to cut plant stems, probably of wild wheat and barley. Pestles and mortars were used to process the grain. The women of some modern sedentary hunter-gatherers maintain kitchen gardens near their homes, tending wild species and using them for food and medicine. It is likely that the Natufian women did the same, particularly cultivating natural stands of wild plants yielding high energy foods such as nut trees, cereals and lentils.

Another early settlement is found in North West Syria on the flood plain of the Euphrates, at a place called Abu Hureyra. Evidence of a small village near a river has been found from 11,500 BCE. These people lived in houses of wooden frames cut into the ground, with reed roofs. Inside they slept on hides and grasses; grinding stones, stone tools, bones and baskets littered the floor. Outside were cooking areas and work spaces. Wild species of plants were probably tended in the surrounding area, weeded and cleared of pests, eventually gathered and brought back to the village, where they were then prepared. Wild gazelle were mainly hunted, but also wild pigs and asses. Evidence of early trade is found in shells from the Mediterranean and obsidian from southern Turkey.

Archaeologists believe that the evidence points to the villagers having stayed put all the year round. The warm, wet climate produced abundant and reliable food sources, and although the people remained hunter-gatherers, not farmers, they became sedentary. They stayed at this village, rebuilding as necessary, for more than a thousand years. The climate then deteriorated with the arrival of the Younger Dryas cold and dry period, and the site could no longer support the people. They therefore moved off, presumably to become mobile again. By 9,000 BCE the site was occupied again, but this time the people were farmers.

It seems that sedentary life came first, and farming followed. As farming spread, the two may well have gone together, but the Neolithic revolution – farming and the formation of towns, leading to the first civilisations and city states – seems to originate not so much in the development of farming, but in the initial decision to settle down.

Living in a permanent settlement has a number of disadvantages. It is a source of social tensions, since people cannot now move away from conflict. It is a source of disease and filth. It must have been inconvenient in many ways as food and other resources near by were depleted. Later as farming and herding were developed, a fresh range of diseases and parasites were communicated to humans from the animals. Farming itself produces a good store of calories, but tends to be nutritionally much poorer than the range of foodstuffs collected by hunter-gatherers. The life itself can be much harder work, and less rewarding, than the hunter-gatherer lifestyle.

A British archaeologist, Vere Gordon Childe, was the first to coin the phrase "Neolithic Revolution". In 1928 he described three transitions worthy of the title revolution in Asia and Europe: the Neolithic Revolution of about 10,000 years ago, characterised by the development of farming; the Urban Revolution of around 5,000 years ago when the first cities developed in Mesopotamia, Egypt and India; and the modern Industrial Revolution. The latter two revolutions arose from the first. Independent Neolithic Revolutions also occurred in Africa, China and the Americas.

According to Childe, people domesticated plants and animals to provide a secure source of food, under pressure from difficult climatic conditions. The production of a steady supply of food allowed a sedentary life style and an escape from "savagery". New technologies then followed from village life and the opportunities it afforded. But the evidence above shows that settling down, or sedentism, came *before* farming.

Thus there are two components to the Neolithic Revolution, the move to sedentism and the development of farming, whereas many theories tended to treat them as one. We have seen that the former began thousands of years before the development of farming in the Near East, and there are signs of near sedentism (where conditions allow it) going back some time before that, at least to the Last Glacial Maximum. The period of growing sedentism and cultural development before the Neolithic is referred to as the Mesolithic.

It was in this period in the Near East that people invented very fine tools for all kinds of purposes, developed house-building technology, and began to ensure a permanent supply of food near the settlement. This eventually led to the domestication of certain plants and later herding animals, and the storage of grains such as wheat and barley, plus peas, lentils and vetches in special deposits within the village. Until then, Mesolithic people continued to hunt and fish and collect wild plants as a major part of their activity.

In 1966 two botanists working in Eastern Turkey noticed that the original wild cereals on the hilly flanks of the Fertile Crescent – emmer wheat, einkorn and barley – still covered thousands of hectares. One of the botanists, JR Harlan, went out with a flint-toothed sickle and within one hour collected enough wheat to produce a kilo of grain – with a protein content twice as high as modern wheat. He worked out that a family could gather more than enough food for a year after three weeks of moderate effort. They asked why on earth anyone should bother to develop farming under such circumstances.

In the 1970s Jacques Cauvin suggested that there had been a symbolic revolution before the Neolithic Revolution, characterised by changes in architecture (from round houses to rectangular, which could be packed together even tighter) and in religion, with the finding of skulls and horns of aurochs embedded in the houses, and also of female figurines which he saw relating to goddess worship.

Cauvin pointed out how the Mesolithic peoples had the sickles, grinding stones and hoe-like wooden tools, along with the abundance of wild cereals, that were pretty well all the people needed to take to agriculture. Why did they not do so until the Neolithic? Cauvin suggested that people had to develop a new mastery of symbolism before they got the idea of mastering the environment instead of passively collecting from it. He saw the Mesolithic as a period of revolutionary development of ideology, religion and psychology, leading to the Neolithic. The veneration of the aurochs was about coming to terms with the dangerous forces, fears and insecurities within us, reflected in the outside world.

His ideas were developed by Ian Hodder, lead archaeologist at the ancient site of Catalhoyuk in Turkey. In 1990 he suggested that people had to domesticate themselves before they could domesticate the environment. He noticed an association between houses, burials and tombs, objects from the wild such as stag antlers and representations of the wild in art form. He suggested that the interior of huts and shelters in the Paleolithic must always have represented warmth and security, and became more and more established, leading to sedentism. It became associated with the social and cultural as opposed to the wild and natural. By bringing symbols of the wild into the house, or within the village, people symbolically domesticated the wild. Having domesticated themselves, and symbolically domesticated the wild, they were ready to actually domesticate their outer environment.

Gobekli Tepe in southern Turkey is the site of an amazing temple complex, the earliest known example of monumental architecture. It consists of rings of large limestone pillars covered in bas-reliefs of animals such as gazelles, snakes, scorpions and wild boars – yet this is currently dated to about 11,600 years ago, before there was farming; possibly even before sedentism for these people. Though they were hunter gatherers, they had the resources to build this permanent monument (and to periodically rebuild). The lead archaeologist, Klaus Schmidt, thinks that this supports Jacques Cauvin's ideas – first there was a revolution in symbols, allowing sophisticated religion, and people then needed to take to farming in order to maintain it.

Michael Balter in *The Goddess and the Bull* suggests that our biological inheritance as a social animal is the key, particularly the revolutionary changes about 40,000 years ago which was marked by an explosion in human symbolism. This allowed for the first time a proliferation in human networks and the ability to maintain close relationships with people generally far away and seldom seen - a "release from proximity".

Modern humans are very different from other social animals, in that the closeness of our emotional ties to each other may not be correlated with geographical closeness. Dearly loved members of our family, and close friends, may live some distance away. We don't forget these people even if they live on the other side of the world, and we look forward to reunions with them. Some of our oldest and dearest friends may actually be those who now live furthest away from us.

This ability to hold the image of certain people in our minds, and metaphorically in our hearts, would have had an important effect in the Ice Age. In harsh conditions, hunter-gatherer groups would have had to be quite small to survive in a given area. However the periodic gatherings that we noted earlier, when several groups came together for weeks at a time to exchange food, tools, ideas and sexual partners, were made possible by this ability for the groups to remember each other as friends and family, not strangers.

Francois Valla, who excavated 'Ain Mallaha, believed that Natufian villages grew simply from the seasonal gatherings of their predecessors. He cited the work of a social anthropologist, Marcel Mauss, who lived with hunter-gatherers in the Arctic at the beginning of the 20th century. Mauss described their periodic gatherings as characterised by feasts and religious ceremonies, intellectual discussion, and lots of sex. The rest of the year, when they lived in small, isolated groups, was rather dull. Valla suggested that people extended their periodic gatherings as conditions allowed, and when the increasingly abundant conditions after the LGM and the beginning of global warming allowed people to stay together all year round, they did so. Gobekli Tepe may reflect a new abundance of natural resources, rather than a major change in consciousness.

Thus we stay together because basically, as a species (there will always be exceptions), we like being together. We may argue and fight and create dirt and smells but the excitement of communal living is preferable to living with just a small family group. Our sociability, which enabled us to live together cooperatively in small wandering tribes, led us to join together on a large scale.

Perhaps it didn't happen before because there had to be a combination of the biological urge, the cultural and social development that enabled us to cope with living on top of each other, the technological developments to do so, and the environmental conditions of abundant resources of food, water and building materials to make it all possible. The initial impetus was almost certainly the periodic gathering of hunter-gatherers that began in the Ice Ages. This probably developed in association with religious and other ceremonies (think of the age of cave art), and gave enjoyment and a wider pool of potential sexual partners as part of the social advantages.

Once we had come together, it would make sense for farming to follow. The first sedentary societies got a rude shock when the Younger Dryas period arrived, with cold dry conditions reducing the abundance of the original Garden of Eden (the word Eden meant the countryside, in ancient Sumer). The settlements now broke up until the climatic blip was over, though there is also evidence that the Natufians were beginning to exhaust their environment. Having got used to a particular way of living, and with all the necessary tools and knowledge in place, perhaps also with a new way of looking at the environment and a new willingness to domesticate it, it may be that farming now developed so that people could continue to live together as they liked.

It was probably women who contributed the most to the development of farming, which is ironic since ultimately in many ways it was not good news for them. Growing crops emerged quite naturally from the gathering of cereal seeds. Seed that didn't fall too easily would be favoured, since it would remain on the plant to be gathered, and could be processed by threshing and grinding back at the village. Eventually the time would come when artificial selection of such seed would mean the plant could no longer drop its seed itself, but depended on human help. It would probably have been women who developed the first wild gardens into a field of crops for maximum harvest. It may have been women who first domesticated the wild dogs that came to scavenge, or the young of hunted pigs and cattle that were brought back to the camp.

It has been suggested that one reason why women may have been the ones to domesticate wild species is their role in nurturing and raising children. They would be used to tending and nurturing to promote survival, and also to the manipulation that goes into the socialisation (training and discipline) of children. Modern gardening and the keeping of pets may give us some insight into domestication and the time and love and care (and intelligence) that has to be expended. It may have been carried out when a woman was required to take care of the children or grandchildren and her mobility was thus more restricted.

The invention of farming was, as we have seen, a major part of the Neolithic Revolution. Production of crops meant that huge amounts of food could be produced and stored, thus supporting much larger populations. With the growth of larger populations and thus larger settlements, the first towns developed. It was no longer an option to go back to hunter-gathering; crops must be produced or large numbers of people would starve. Often through history, as soils were exhausted or rains failed or pests decimated the crops, the people did starve.

High calorific value meant that populations exploded, but at the same time nutritious content fell and health declined. Depending on a few plant species meant that essential vitamins and minerals were often lacking. Living in close proximity to each other and to animals, with little hygiene, meant that diseases and parasites were rife. Doing hard, repetitive work such as grinding corn caused marked deformation of bones that can be seen in archaeological specimens today. There was evidently high infant mortality and people were of smaller stature.

There is evidence that children were weaned earlier. This was possible because of the use of animal milk and mashed cereals, and meant that women could get back to work quicker. This contributed to the population explosion. It also meant a further loss of freedom for the women and their yoking to endless back breaking labour.

It is not clear exactly how class society arose. It seems likely from Christopher Boehm's work on modern hunter-gatherers that the role of any leader among our ancestors was to ensure security and prosperity for the tribe, rather than exploitation. Anyone who tries to lord it over the others and gain special privileges tends to be penalized.

There would normally have been very little to exploit anyway in a hunter-gatherer community. However, Gobekli Tepe shows that even before the Neolithic revolution a big enough surplus was produced such that the community must have grown considerably, and could support people who were not directly productive themselves. This may initially have freed up religious work. It might have included leaders to oversee various projects. It would become increasingly difficult to prevent a hierarchy of roles.

Eventually someone must have had the idea of laying claim to the surplus, or at least the management of the surplus, and used it to maintain a band of supporters who would back them up in return for material favours. This sort of behaviour might have its roots in the sort of chimpanzee behaviour we saw earlier, of allies banding together for mutual gain. The position of exploiter could then be maintained ultimately by brute force, though no doubt it was from the first backed up with religious and ideological justifications. Kings were supposed to be chosen by the gods/goddesses.

The beginnings of agriculture seem to date back as far as about 9,000 BCE in the Near East. By 7,000-8,000 BCE in the Near East there existed well-established towns such as Jericho and slightly later, Catalhoyuk in Anatolia with as many as 1,000 houses and 5,000 people. By 3,000 BCE the first cities had arisen in ancient Sumer (now southern Iraq), and with the invention of writing came the end of prehistory and the beginning of history.

Sumer was centred on the banks and tributaries of the Tigris and the Euphrates, from about where Baghdad is today down to the Arabian Gulf. Silt was carried down by the rivers to cover the flood plains, providing fertile soil for intensive agriculture. Sumerians called their country "the land". It was made up of a number of city states, such as Eridu, Ur, Lagash, Uruk and Nippur. Lagash is said to have eventually numbered about 30,000-35,000 people. Each city and its territory was under the protection of the particular god/goddess to whom it belonged. Records show that about one-third of the land was owned by the temples, which employed thousands of people including a large bureaucracy and which also had slaves.

The palace employed large numbers on its estates, probably as large as the temple estates, and kept up an army – one example talks of 600 or 700 soldiers with their equipment. The king governed the city state on behalf of the gods. The rest of the land belonged to freemen of varying wealth and status. Everything was tightly controlled and most people were due to be called up for army service as required. One of the duties of the king was to maintain the temples, following the Sumerian belief that people had been created simply for the service of the gods.

The class system was well under way by this time, and we know details of its operation because of the many written tablets that have been found by archaeologists. We know that in the Roman Empire about two-thirds of the peasant's crop went to support the ruling class, leaving them with a subsistence income – or less if things went wrong (drought, disease etc). This sort of arrangement probably first became common once a particular town was large enough that its ruling class could dominate all the countryside and farms in the district around it.

With the rise of the class system came the domination of women. Although there have been examples of ruling women through the ages, generally speaking it is the men who were the first rulers, largely because until comparatively recently rulers had to prove themselves on the battlefield. Also women were bound by biology in terms of pregnancy, childbirth, nursing etc. There are arguments today as to just how soon women were subjugated, but it seems likely that class and patriarchy went hand in hand.

Eleanor Leacock for example showed how the position of women in Montagnais-Naskapi tribe in northern Canada changed as their tribal mode of production was altered by the arrival of European, early capitalist society. When the Montagnais-Naskapi were hunter-gatherers, the economic unit was the whole tribe and men lived with their wives' families. Although men and women had different roles, these were quite fluid and women could have power and influence. As production changed to trading for profit with European settlers, men took on the role of main breadwinner and head of a privatised family. Women had now lost their power.

So what happened to the Grandmother and the Wise Woman, archetypal roles which changed the whole life course of the human female and gave us the chance of a long life after menopause? The grandmother role has remained of course, though it may never have been as important after the invention of agriculture. As for the role of the Wise Woman, the Neolithic revolution changed that out of all recognition.

The key to success in Paleolithic times was knowledge, as each band had to know where to find the huge number of plants, animals, water holes and other resources of the landscape that would enable their survival. Once the Neolithic revolution had occurred, the key to success was labour. Endless, back breaking labour: toiling in the fields, building dams and irrigation systems, threshing the harvest and grinding the grain. Knowledge became the privilege of the ruling class, the kings and major landowners, the priests and the leading bureaucrats. The surplus produced by agriculture made a division of labour possible, and most people – male and female – were on the wrong side of the division.

Of course the Wise Woman role remained for some women. We shall see later that women could be powerful as priestesses, and are known in some places to have acted as women Elders. There seems to have been some variation in the degree of patriarchy in early civilizations, and some argue that where the Goddess was worshipped, the status of women was high, at least among the ruling class.

Among the lower orders older women still had some useful knowledge, for example in the knowledge of herbs and their uses, particularly those to do with reproduction. But the old power and respect were gone, and such women were sometimes subject to persecution, as with the witch trials.

It is too late now to turn back the clock. There are now simply so many people in the world that we must have agriculture to feed them, and modern intensive agriculture to feed them well. Anyway we have lost the knowledge and the appropriate landscapes to exist as hunter-gatherers. And I doubt many would be willing to give up the comforts and medical care of modern civilisation.

Women have lost the particular role that brought about the evolution of the menopause. What we have gained is a world where at least in its developed areas, we have the potential to take on any of a huge range of roles according to our sense of what is right for us as individuals. That potential needs to be developed, not just for us in the West but all over the world. We can't go backwards, but we could work hard at laying the ground for future developments.

It can be an interesting exercise to contemplate just how long our role as Wise Women endured, compared to the short time that patriarchy has been in existence. Take a tape measure or a ruler and measure out two metres. Allowing each millimetre to represent 1,000 years, then two metres covers 2 million years. Mark near the beginning of the length, nearly 2 million years ago, the rise of early humans – the hunter-gatherer *Homo erectus*. At 60 centimetres, mark the beginnings of archaic *Homo sapiens.* At 5 centimetres, we have the beginning of the Upper Paleolithic and the explosions of art and symbolism about 50,000 years ago. At one centimetre, mark the Neolithic Revolution, the beginning of farming and the decline in the role of the Wise Woman. Mark the rise of city states about 5 millimetres before the end of the line, and of modern capitalism just about at the end of the line.

Civilisation has been here for such a short time, in geological terms just the blink of an eye, and it has completely transformed the whole world. Not just our social world, but the ecosystems of the entire planet. Human activity has obliterated species, cut down vast forests, dried up lakes and created dust bowls. Now we find that the Industrial Revolution has caused global warming on a scale that has serious implications for our current civilisation's ability to survive, if we don't take swift action to change our ways.

Our role as Wise Women was of key importance for millennia, and perhaps our wisdom is in fact *more* urgently needed today than ever before. Older women have traditionally been the carers, the nurturers of the tribe, as well as the holders of wisdom, and the ones who first tended the fruits of the environment. Perhaps the wisdom needed today is to see beyond our immediate concerns and view the world as a whole, where the activities of one set of people on one side of the world can have a devastating effect on the peoples living in a completely different part of the world.

We looked earlier at the concept of the "release from proximity", a development in human capability which meant that we could care about those who were not immediately before us. Perhaps the role of the Wise Woman today is to develop this capability to care for the world as a whole, and all the peoples in it. No one person can change the world, but together as we each do our little bit, we can make a difference.

CHAPTER 9 – The Great Goddess

We tend to take for granted today, in our mainly Judeo-Christian-Muslim western world, that God is male. He is the Father that sits on the throne of heaven, possibly opposed by another male supernatural being called the Devil who rules down below in Hell. His Son in Christianity is male; the Holy Spirit is generally spoken of as male. His prophets and priests have generally been male.

From this point of view, women are often the problem. Women represent sin – from Eve tempting Adam in the Fall, to the Whore of Babylon and all the imagery of lust and sinful sex. Men fall because women tempt them. When they are not sinful, they may be seen as simply weak and irrelevant and required to submit to male authority. The various female bodily functions may even contaminate the church. Anything to do with menstruation and childbirth may need careful regulations and rituals. The main female role model in Christianity is the Virgin Mary – but who can emulate a virgin birth?

There are contradictions in this view of women. On the one hand they are pathetic weak creatures who need to be guided and ruled, on the other hand they have a dangerous frightening power that must be suppressed. Neither view tends to be stated so bluntly nowadays apart from in some fundamentalist churches, but the undercurrents are still there. Neither view is very helpful to the woman entering the third phase of her life and assessing her identity and her role.

Given the way that patriarchal civilisation has suppressed the role of women Elders, we might expect to find that the same thing has happened in religion. In fact worship of goddesses (and other pagan deities) was finally outlawed by the Emperor Theodosius in 394 CE (Common Era, or AD). Although the position of ordinary women may have suffered under class society, goddess worship was once common even in the West. And if you google "goddess" today, you will come up with millions of sites. As women have sought to throw off the chains of patriarchy in the last few decades, many have also reclaimed the Goddess.

Today many women believe that for much of the span of time from the final evolution of fully modern humans and the mental capacity for religion (perhaps about 40,000 years ago, though it may have evolved much earlier) until relatively recently, goddess worship was the norm. They believe it was the arrival of patriarchal tribes from the north that brought about its suppression in Europe and western Asia during the period 4,500-2,500 BCE.

This idea owes much to the work of archaeologist Marija Gimbutas. Gimbutas suggested a golden age where women were not oppressed, the Great Goddess or Goddesses were worshipped, and where fertility and sexuality were celebrated. The societies that worshipped the Goddess were communal, peaceful places where women had high status. This lasted until warlike patriarchal tribes descended and imposed their own gods and values including the rule of men.

Long before Gimbutas' work, Western scholars knew that female deities were worshipped in other societies such as India, and that goddess worship had preceded Christianity in Europe. In 1861 J J Bachofen published his book *Mutterecht (Mother Right)*. This was based on the idea that societies evolved from a primitive state through various stages to a more advanced and civilised state (in other words, to the modern imperialist Europe of his time, which he saw as the ideal). It was comparable to a child growing up to be an adult. People from still existing hunter-gatherer societies were considered to be childlike savages, inferior to White European males.

Bachofen suggested, on the basis of old myths, that originally societies were matriarchal – power was held by women. This arose from the close bond between mother and child, and the mother's authority. However this was a primitive stage and later evolved into patriarchy – power held by men – which was, of course, a superior arrangement.

An evolutionary view of social development was influential for many people at the time, being held for example by Lewis Henry Morgan, who studied matrilineal descent (descent traced through the mother) in native American tribes in the 1870s. In his book *Ancient Society* (published 1877) he developed the idea of social evolution, linking social change with technological change. These ideas were taken up by Friedrich Engels, to show how social relations depend on the mode of social production.

Sir James Fraser was another who looked at old myths, gathering a huge collection of myths into his book *The Golden Bough* (1911-15). These myths focused on the mother goddess and her male son/consort. In these myths, a female goddess would hold supreme supernatural power. She would then enter into a sacred marriage with her son (or sometimes her brother) and conceive by him before he died, or he would die in the autumn and be resurrected in the spring. This reflected the eternal cycle of the seasons of growth, death and then rebirth.

Freudian readings of the myth stressed the power of the mother over the infant, she being the source of all things, good and bad. There is also the son's desire for her and the developing male's need to break away from her influence and to reject the female world for the male. Jung saw the Great Mother as an eternal archetype. He compared the mother to the unconscious, and the paternal principle (consciousness) has to struggle for supremacy. The 1955 book *The Great Mother* by Erich Neumann was one of several influential books combining archaeology with psychology along these lines.

Not all saw patriarchy as a superior stage. In 1891 Elizabeth Cady Stanton, an American feminist organising for women's right to vote, made a speech entitled "The Matriarchate, or Mother-Age" concerning goddesses and ancient matriarchal civilisations. Some American feminists were aware of the matrifocal societies (societies where descent was traced through the mother and women's status was quite high), still surviving in their own country among the native Americans. Cady Stanton declared that knowing their own history of powerful women could give women a new sense of dignity and self respect.

In 1967 James Mellaart wrote of his excavations at Catalhoyuk in Turkey, dated from about 6000 BCE, and was greatly influenced by the ideas about the Great Mother Goddess of early societies. His interpretation of his findings as of goddess figurines, temples and an egalitarian society reflected this. Excavations of early Neolithic Mureybet (Syria - from around 9000 BCE) by Jacques Cauvin in 1971 revealed clay female figurines, on the basis of which Cauvin proposed a cult of the Great Mother Goddess.

In 1974 Marija Gimbutas published her book, *The Gods and Goddesses of Old Europe* and became immensely influential on a burgeoning Goddess movement. (Her book, now entitled *The Goddesses and Gods of Old Europe,* was originally titled *Gods and Goddesses* on her publisher's orders.) An archaeologist, Gimbutas took the evidence to show that there had been a peaceful, egalitarian Goddess-centred culture in 'Old Europe', and also in western Asia.

Gimbutas led digs at a number of Neolithic sites in southeast Europe (the Balkans, Greece, Italy) and interpreted her finds as showing an 'Old European' culture that was goddess-centred and matrilineal. She found a range of female deities: snake, bird and bee goddesses, frog, fish and butterfly goddesses, Mistress of the Animals, bear mother, craft-giver, life-giver, birth-giver, nurse, earth mother, goddess of death and more. She saw these as manifestations of the Great Goddess in her many aspects. She thought that these societies, while not matriarchal (ruled by women), were more egalitarian than later patriarchal societies so she referred to them as matrifocal or matristic.

Gimbutas' work was popularised by books such as *The Paradise Papers* by Merlin Stone, published in 1976 (entitled *When God was a Woman* in the USA), and *The Chalice and the Blade* by Riane Eisler in 1987. The influence of these ideas is strong, but many archaeologists have questioned the ideology that has grown up around the search for the Great Goddess, and regard it as a modern myth. Literally true or not, the myth has great resonance for us now as we reclaim our own sense of power as individual women.

Archaeologists have tended to emphasise the Great Goddess as a Mother Goddess. In fact her title in the Near East was usually 'Queen of Heaven', and her roles included creator, lawmaker, warrior, hunter, wise prophet, and goddess of the various trades and professions that made up society as well as of agriculture and fertility. In many societies she was seen as a sun goddess. Her societies were originally seen by many male archaeologists and anthropologists as dark, primitive and chaotic, with civilisation arising through god-worshipping cultures such as classical Greece. But civilisation, in terms of great towns, written language, laws, agriculture, government, medicine, crafts and trades, was initially developed at least in some areas by people who were worshipping goddesses.

The Goddess bore many different names in different towns and eras, although some argue that they were essentially the same Great Goddess – as we could argue that Yahweh, God and Allah are essentially the same deity today. In Canaan she was Astarte, Ashtoreth, Asherah; in Mesopotamia she was Inanna (Sumer) and Ishtar (Babylon); in Egypt she was Isis, Hathor, Au Set; in Syria she was Anat, Athar; in Turkey Arinna. She can be identified with Brigid in Ireland and Shakti in India.

Why would the first god be a goddess? One theory goes that early peoples did not connect sexual intercourse to childbirth, so it seemed that women only were involved in the great mystery of birth. Given the high level of observation and knowledge of the natural world among hunter-gatherer societies, this seems unlikely, and in fact few proponents of the Great Goddess idea hold to that theory. Certainly females were the ones that carried the child, gave birth and then suckled their infant. It was very clear who the mother of the child was, and not so easy to say who the father was. Modern surveys such as that of Robin Baker and Mark Bellis suggest that some 10% of children are fathered by a man other than the one supposed to be their father.

The child was born to the mother and thus to the mother's clan. This way of tracing descent, through the female line, is known as matriliny or matrilineal descent. What generally happens (from observing modern matrilineal peoples) is that after marriage the man goes to live with the woman's relatives. Men still often hold the more powerful position, but trace their power through the woman – it may be the brother or uncle who rules. The status of women is generally higher than in patrilineal societies.

As well as ideas of descent and inheritance, the role of women in giving birth would have given them powerful associations with ideas of fertility and abundance. Fertility and fruitfulness was both a sacred mystery and a vital necessity to our ancestors and the female was seen as key to this. The bringing forth of human babies mirrored the productivity of the plant and animal kingdom, without which we would have starved. The power of women to have babies, a power which men lacked, was acknowledged and revered. The power of life was coupled with the mystery of death so that that too was linked to the power of the Goddess. After we had emerged from our mother's womb and lived out our life, we would return to the earth – the womb of the Goddess – to await rebirth. Bodies were often buried in the foetal position.

Blood is the life force and yet women shed blood regularly in their menses without any harm to themselves. Childbirth, an awesome thing in itself, also involved blood. It is likely there were always plenty of taboos and superstitions and segregation around the question of women's blood, but originally it would be with a sense of a sacred though potentially dangerous power. Later on under a monotheistic male god, there was still a sense of mystery and a sense of women's power, but it became more negative. Menstrual blood was to be prevented from contaminating men. Sexually aware women became regarded as whores and she-devils, although originally sexuality in women could have been something to celebrate and revere.

In the Neolithic period people were still working with stone rather than metal, but turned from the old hunter-gatherer existence to settled villages and farming. Women may have been the ones to develop agriculture and the domestication of wild plant species, since they were the ones most likely to have been in charge of gathering plants and tending the first wild gardens. This power over nature might also have contributed to the power attributed to the Great Goddess.

Ancestor worship seems to be one of the earliest characteristics of religion, and as women were the known bearers of children, it could have been a woman who was the First Ancestor, the progenitor of the tribe – as is found in many myths.

Thus the Great Goddess would be the one with the power of creation, the one who created the tribe and all its world, the one who would continue to create abundance in the fields in terms of plant and animal life (or to withhold it in bad times). She would be the one who would confer legitimacy upon earthly rulers. She was the source of all things, and when people died, they would return to her womb.

According to Gimbutas and her followers the less hierarchical, more peaceful times of the goddess were interrupted by waves of invasions of Aryan/Indo-European peoples from the north, certainly underway by 2,400 BCE and possibly starting as early as 4 or 3,000 BCE (the Kurgan hypothesis). These tribes may have come from the forest and coastal areas of Northern Europe, Russia and the Caucasus, originally developing their culture in the period 15,000-8,000 BCE. They were pastoral people, perhaps originally nomadic hunting and fishing peoples, later herding animals.

These people were war-like, in particular introducing the horse-drawn war chariot which gave them a tremendous advantage over more civilised and more peaceful societies. They invaded and conquered other territories in waves of migration, and also fought each other. It has been suggested that perhaps the living was not so easy for them in Northern Europe than it was for people in the Near East, and there was fighting over scarce resources. The northern tribes expanded and conquered to the west (Western Europe), south (Greece) and east (Central Asia, Northern India, Iran and Turkey). They entered Turkey, Iraq and Iran from around 2,400 BCE, then into Syria about 1800 BCE and Canaan 1500 BCE. Later they became the Germans and the Celts, among other Indo European types.

This was a patrilineal society. This means that descent was traced through father to son, not mother to daughter. Why patriliny took hold is not known. Perhaps a predominance of hunting and herding in the original Indo-European tribes meant a woman's role and therefore her status was less valued. Certainly once the tribe was run by warriors then patriliny would seem more likely. The tribes had their own supreme male god with a priestly (male) caste and saw themselves as superior to the people they conquered, even when they adopted aspects of civilisations more advanced than their own.

These tribes brought their ruling male deities and institution of kingship, with the emphasis on God as King. They tended to have storm gods, holy mountains, an emphasis on the light of fire or lightning. They brought the dualities of light (good) and dark (bad – was this partly because of their fairer skins?). In their later myths their own god married or destroyed the goddess of the indigenous people they conquered. There were many myths whereby the old goddess was killed in the form of a dragon or serpent which often carried a name suggestive of the goddess (the goddess was often associated with snakes).

Gimbutas found that when the northerners conquered the matrifocal societies the result was a fusion of the two different systems. As time went on the patriarchal elements – including male gods - prevailed. The Indo-European conquerors immediately adopted the art of writing from their conquered peoples. This means that most of the written myths we now have are already male-dominated, and in these myths the male gods gradually take over. However goddess worship still survived. There was usually a panoply of gods and goddesses, and although the most powerful god was now male, female goddesses still held divine power of their own.

Some of the earliest writings referred to Inanna, the Queen of Heaven, in Sumer (modern day Iraq) some 5,000 years ago. By this time, according to Gimbutas, the Indo-Europeans had already begun their conquests, and most of written history has been composed by the patriarchal male conquerors. How can we find out what was happening before the invention of written language, and before writing was used to record myths? This is where both myth and archaeology are important. Studying myth can give us pointers to how they developed, while archaeology provides the concrete evidence such as figurines, carvings and the remains of temples – though both are still open to interpretation.

CHAPTER 10 - Development of the Great Goddess

There are numerous sculptures of women from the Upper Palaeolithic onwards (from about 25,000 BCE). Probably the most famous of these is the Venus of Willendorf. These small figurines are made of stone and bone and clay, and usually show women with large breasts, stomachs and buttocks, perhaps pregnant. It has been suggested that the fat distribution in some of the figurines is actually showing a menopausal woman. They are similar to later Neolithic figures of Goddesses, which suggests they were made for similar reasons. They are found all over Europe, Russia, Canaan, Turkey, Iraq, Syria.

Supporters of the Great Goddess theory point to evidence such as Neolithic Jericho (about 7,000 BCE). People were already living in plastered brick houses. Rectangular plaster shrines were found and female clay figurines resembling idols of the Great Goddess. At Catalhoyuk (in modern day Anatolia in Turkey) dating from about 6,500 BCE, about 40 shrines were found. The principal deity was a goddess in three aspects: young, a woman giving birth, and old. Female burials were associated with red ochre. In the Halaf culture in Near and Middle East in 5,000 BCE there were towns, metallurgy and wheeled vehicles. Goddess figurines were associated with representations of serpents, double axes and doves. In Eqypt around 4,000 BCE the emergence of agriculture is associated with the Goddess figurines.

Goddess supporters suggest that in early Neolithic art there is an absence of scenes of battle, slaves, or the conquered peoples being driven in chains. There appear to be no pictures of noble warriors, and an absence of lavish chieftain burials. No caches of weapons have been found, simply tools for everyday life. Fortifications do not appear until later, apart from walls built to keep out floods rather than people. They say that at Catalhoyuk the houses are the same size and the burials are similar, which suggests an egalitarian society. Neither male nor female dominated; hierarchy and ranking were not relevant. Power was about responsibility, not oppression.

Male as well as female elements were present, for example the horns of wild bulls widely found at Catalhoyuk and other places, as well as phalluses. The two elements were harmoniously linked in the yearly ritual sacred marriage of the Goddess and her young son/lover/husband. This probably involved the chief priestess as the representative of the Goddess and therefore queen of her people.

Proponents of the Great Goddess see Neolithic art as expressing the abundance of nature and its essentially benign aspect, although the destructive forces of nature are also recognised and respected. The primary purpose of art and life is to cultivate the earth and respect the Goddess, not to conquer and pillage and exploit. No doubt there were times when this was not adhered to, but domination was not seen as the ideal.

Prehistory becomes history with the invention of writing, around 3,000 BCE. The first writing found involved the Great Goddess - just before 3,000 BCE at the temple of the Queen of Heaven in Erech in Sumer, the business accounts of the temple are written in cuneiform script. The Goddess is found in written myth in all areas of the Near and Middle East, until the last temples were closed by Christian authorities in around 500 CE.

The names of the Goddess vary in different areas, but tend to be simply titles of the Great Goddess such as Queen of Heaven. Symbols associated with her include the serpent, cow, dove, double axe. She undergoes an annual sacred marriage with a son/lover who (symbolically) dies. Under her sway, female priestesses rule the temple assisted by eunuch priests; the sexual act is a holy one as the source of life and is undertaken by priestesses with men visiting the temple to honour the goddess. Women of the temple were known as holy women and were often women of property and businesses. Children born to the priestesses inherited temple property and were legitimate and respectable. Who fathered them was of no importance.

It may be that the sex of the main deity is determined by the sex of those in power, mirroring the gender of the head of the family. Thus the influx of male ruling deities mirrors the patriarchal nature of the invading tribes. Did Goddess worship ever mean there was true matriarchy, rule by women? Matrilineal descent does not mean matriarchy, but the status of women under goddess worship was much higher than under patriliny and supreme male gods.

Merlin Stone quotes Diodorus of Sicily in 49 BCE who commented on the high status of women in some countries he visited such as Ethiopia and Libya. In spite of years of male supremacy, goddess worship still seemed to be associated with something more approaching equality. In Egypt Isis was revered as the inventor of agriculture, physician, lawgiver. Women had higher status there, and it was the daughter who inherited the throne so her brother married her in order to become pharaoh.

The temple of the Goddess was an important centre of agriculture, trade and business, and had great influence over even male rulers. One tablet in Sumer said of Agade, where King Sargon lived: "Holy Inanna established the sanctuary of Agade as her celebrated women's domain… She endowed its old women with the gift of giving counsel, she endowed its old men with the gift of eloquence" (ETCSL 2006). The Hittites had a group called The Elderly Women – having a prophetic, advisory position and involved with both mental and physical healing. The earliest Sumerian myths had both female and male deities in decision-making assemblies in heaven. Could this have reflected the way the society producing those myths ordered their government? Was there community government? Later monotheism would reflect the supreme power of the king.

Some believe that the myth of the yearly sacred marriage reflects what actually used to happen when the Goddess was supreme. The high priestess as the incarnation of the goddess would take a young lover in an annual ceremony of sacred marriage or sexual union. This consort would now be king for a year. He would then be ritually sacrificed in the autumn, to be 'resurrected' (the next marriage) in the spring. Or the consort would be killed when he was no longer willing to defer to the Goddess/priestess. In the legend of Inanna and her descent to the Land of the Dead, when she had to find a replacement for herself to remain there, she passed over those respectful to her and chose her consort Dumuzi because he had joyfully taken over her throne in her absence and then acted arrogantly on her return.

It is suggested that when the northerners took over kingship, they made it permanent. Probably at first the kings took the high priestesses in marriage to confer legitimacy on their reign. The Kings of Sumer were known as the 'beloved husbands of Inanna'. Later as the kings became more powerful they made their own relatives priestesses. In the ancient legend of Gilgamish, he was an early En (priest of the Goddess) of Erech – a city in Sumer. Gilgamish refused to marry the goddess Ishtar on the grounds that he would end up like her previous consorts – dead. Gilgamish kills the Goddess's sacred bull and throws its thigh bone (or possibly its genitals) in her face, saying he would do the same to her if he could. Gilgamish then goes off to be a hero.

In Babylonian times, the King was ritually humiliated in a New Year festival, and a substitute was sacrificed. This eventually represented making amends for the sins of the people, but may originally have come from the old fate of the consort. The cult of the Great Goddess Cybele from Anatolia was transplanted to Rome and celebrated up until 268 CE. In ritual reenactment of the fate of her consort the young shepherd Attis, an effigy of Attis was tied to a tree and then buried. Three days later he rose from the dead (with a light appearing in his tomb), bringing salvation to the people.

According to the Kurgan hypothesis, as the northern ways became dominant the position of women changed. The father grows in authority, the woman loses the right to choose her own partner, the children belong to the clan of the father, the wife is subject to her husband. This happened at different timescales and to different degrees in different places.

In ancient Mesopotamia, around the Fertile Crescent in the Middle East, various regions vied for supremacy over the years. The first civilisation with writing, Sumer, worshipped the goddess Inanna, a supreme goddess who killed her husband Dumuzi for his lack of respect for her. In the Akkadian era, Inanna was a Great Goddess who gave up her royal sceptre to the god Enki but still had some powers allowed by him. Later still Babylon became supreme. Their male god, Marduk son of Enki, mythically murdered the Creator Goddess Tiamat. Inanna became Ishtar, who now grieved over the accidental death of her husband Tammuz.

We can see the changing position reflected in the codes of law. Under Hammurabi's laws in Babylon (he reigned about 1792-50 BCE), women could still inherit property, conduct business, and were free to request divorce. There were female judges and magistrates and priestesses were providing military and political advice. By 1000 BCE, married women were not allowed to engage in business, unless directed by her husband, son or brother-in-law.

Crete from the Neolithic to the time of northern invasion was said to be matrilineal, Goddess worshipping and possibly matriarchal. Priestesses were powerful and women had sexual freedom. A similar culture was found with the Mycenaeans in Greece around 1350-1100 BC. However there followed waves of northern invasions – the Achaean in the 13th century BC, and the Dorians in the 10th century – both patrilineal. They brought their Indo-European gods including

Dyaus-Pitar, meaning God-Father. Thus we get Zeus in Greece and Jupiter in Rome.

One of the original Greek Great Goddesses, Hera, now became the wife of Zeus, often fretfully resentful of him but subject to his authority and greater power. In 500 BCE in Greece powerful goddesses remained (Athene, Artemis), and their priestesses, but women had become men's possessions. Even the militaristic, patriarchal Roman Empire had powerful goddesses and their priestesses though the sky god Jupiter was the King of the Gods.

In India, the dark skinned Dravidians were a Goddess worshipping people. They were conquered by the lighter-skinned Indo-Aryans in around 1500-1200 BC, who have been described as barbarians compared to the indigenous civilisation. The Indo-Aryan mythology had the A-Dityas (good) pitched against the Danavas or Dityas (bad). From the A-Dityas the god Indra kills the goddess Danu and her son as serpent demons, or cow and calf. The A-Dityas had an ancestral father god, Prajapati, and also Dyaus Pitar. They started the caste system separating the light-skinned Aryans from the dark-skinned Dravidians. The ultimate god was Brahma – 'he whose form is light'. The Aryans spread south but never conquered so completely in the south, so that eventually southern religion became incorporated into the Brahmanic literature.

The patriarchal, patrilineal societies which adopted Christianity were not only influenced by the Bible's ideas about the worth and status of women but also shaped that religion to reflect society's values. Many of the assumptions and prejudices and oppression that women suffer today were originally backed up by the authority of Holy Writ.

Today many women in the Western world take the academic research of Marija Gimbutas, or rather the popularizations of her ideas, as a basis for goddess worship today. They see the idea of peaceful, non-hierarchical goddess-worshipping societies in the past as something we could aspire to return to today. Goddess worship becomes part of our struggle against patriarchy.

The ideology of the Goddess often means that too much is invested in this version of history for an objective review of current research. But does that matter? Clearly much of what Gimbutas had to say was backed up by evidence – and again and again the ancient myths show us male gods overthrowing the power of the goddess.

Lillian Doherty is Associate Professor of Classics at the University of Maryland. In her book *Gender and the Interpretation of Classical Myth* she looks at a number of different ways of seeing or using myth. She suggests that the feminist spirituality movement is using Gimbutas' version of the Great Goddess as a 'charter myth' – justifying a particular viewpoint of the present by claiming that that was the way it was in the past. In other words, some seek to bolster the position of women now by pointing to a golden age when they were powerful both in heaven and on earth.

The psychologist Jean Shinoda Bolen in her book *Goddesses in Everywoman* sees classical myths as a source of validation to her female patients, and says that they are "good to think with". The myths are sources of female archetypes, role models that resonate deep in our psyches and which can inspire us to greater strengths.

The myth of the Great Goddess, as a charter myth, gives us a sense of power in times when women are second class citizens. Charter myths are common because they serve a real, psychological purpose. When a particular group of people feels powerless, "knowing" that things were not always thus can give them the inspiration and courage to keep going or even to fight back.

The idea of the Great Goddess in all her different guises is an immensely useful one as an archetype for women today. And in one sense it is true; we do all carry all aspects of the Great Goddess within us, since ultimately she was made by humans in their own image. Does it then matter if it is not literally true that through all of human prehistory, we worshipped Her and Her alone? and that in such societies the position of women was much more favourable?

My own position is that I want to know the truth, even if it is painful. This is our history. As with conscious ageing, I believe we grow by embracing the truth of our position rather than hiding it. If the Goddess is a real archetype for us, there is nothing to fear in clarity. If the Neolithic revolution brought about class society and patriarchy, how could there be non-hierarchical societies? How do we explain the fact that goddess worship also occurred in patriarchal, militaristic societies? Was goddess worship ever truly universal – how do we know that? We examine the evidence in the next chapter. Even if it turns out that the truth is rather more complicated, we can still use the Goddess in all her forms as our most powerful archetype.

There is such a variety of images that we can choose the particular archetypal image that we need at a particular moment. Enheduanna, daughter of the Akkadian king Sargon and chief priestess, is the first author in the world to be known by name. She wrote three hymns to Inanna which survive and which give us a picture of the diversity to be found in one goddess. The first hymn celebrates Inanna as a ferocious warrior goddess; the second celebrates Inanna's role as goddess of civilisation and the home and family. The third addresses Inanna as a personal goddess who can be directly called upon for help, as Enheduanna asks Inanna to help her regain her position as priestess of the temple against her rivals. Clearly Inanna was a powerful goddess with a diverse range of roles, and even though the country was ruled by a king, women held positions of power within it.

To take Enheduanna's hymns as a simple example, perhaps sometimes we need the image of a viciously ferocious warrior Goddess to stir up our courage and resolve, and to blot out the image of the gentle little woman stereotype. And at another time, perhaps going for a promotion in a male-dominated workplace, the idea of the Goddess of civilisation will help us overcome the internalised idea of men being the ones who run society.

At other times perhaps we would want to use images other than the Great Goddess, as for example animal images were common in prehistoric art. For more detail we could look to the many works on American First Nation peoples, or Celtic pagan ideas, about shapeshifting and the different qualities that each animal may represent. Many "New Age" writers such as Gill Edwards draw on these ideas. We see it in everyday language such as cunning as a fox, wise as an owl etc.

If we are clear about the difference between literal and metaphorical truth, then we can still play with metaphorical truths and paradox that cannot be put into words, but only expressed and experienced as image, symbol and archetype.

CHAPTER 11 – The Evidence for the Great Goddess

Archeologists were refuting some of Gimbutas' ideas even before her book *Gods and Goddesses of Old Europe* was published. Peter Ucko published *Anthropomorphic Figurines of Predynastic Egypt and Neolithic Crete* in 1968, in which he challenged Gimbutas' interpretations. She had seen most of the various excavated figurines as female, where he saw many of them as sexless, without defining features such as breasts or beards. He also challenged the idea that female figurines were necessarily goddesses, suggesting that they could have had any number of purposes including children's toys or sympathetic magic. He pointed out other assumptions, such as that the early Egyptian figurines of women holding their breasts signified maternity or fertility, whereas written texts showed that this was the female sign of grief.

Because early scholarship had assumed that prehistoric people had worshipped the Great Goddess, any human representation that wasn't clearly male was taken to be female, while all sorts of signs such as circles, pits and cups were interpreted as female without any evidence for gender. In 1969 Andrew Fleming in *World Archaeology* challenged the interpretation of spirals, circles and dots as eyes, or that eyes would necessarily symbolise female eyes and in turn symbolise the Goddess.

Archaeologist Timothy Taylor (in *The Prehistory of Sex*) refers to Gimbutas' followers as "misguided", and disagrees with her original evidence from the early Neolithic in Europe, which suggested a peaceful society under the Goddess. He maintains that the people of the early Neolithic had a "fundamentally exploitative attitude toward everything, including sex – being violent, unbalanced people, whose idea of a good time was felling trees, erecting great stone phalluses, and sanctifying them with sacrificial victims, often women and children."

Timothy Taylor emphasises the newly significant opposites of wild/domesticated, inside/outside, male/female, as contributing to the exploitation of both nature and people. As one example he tells of a mass grave found on the edge of an early farming village in Germany. It contains both adults and children, killed by stone axes and cudgels, and two by arrows. He believes this was a raid for slave labour on the new farms. Taylor also believes gender differences in power extend back at least to the Upper Paleolithic, and that the Venus figurines actually represent the objectification and possession of women by men.

Steven Mithen, Professor of Early Prehistory, has written a comprehensive view of global human prehistory in his book *After the Ice: a Global Human History 20,000-5,000 BC*. This examines the broad sweep of the change from Paleolithic to Mesolithic to Neolithic, the giving up of the old hunter-gatherer way of life and the settling down in fixed settlements with the development of farming – or in some cases, a more nomadic way of life by herding. However the details of that story show a marked variation between times and between places.

One example of the early communities in the near and middle east, a Natufian settlement on the Euphrates, shows no signs of conflict. Around 9,000 BCE these people were living off at least 157 different plant species, and probably off 100 more. The area was rich enough for them to settle down and the bones indicate a healthy, peaceable existence.

However in Northern Europe during the Mesolithic period there is evidence that these hunter-gatherers fought aggressively with each other, as Taylor suggested. Excavation of graves indicates that violence was endemic thoughout. There are fractured skulls and embedded flint arrowheads. The majority of the head injuries show blows to the front and left, as if face to face with a right-handed opponent. A cave in Germany shows two pits from 6400 BC, one with 27 heads and one with six heads, mainly from women and children. The male heads in particular show head wounds – one had six or seven massive axe blows.

The suggestion is that the growth of violence in northern Europe was due to population pressure on diminishing resources. As global warming after the last Ice Age led to rising sea levels, the most productive areas next to rivers, lakes and the sea were flooded out. On top of that, farmers arrived in the north from the south east by 5500 BCE which led to social disruption and economic stress among the hunter-gatherers. Mithen takes a less pessimistic view than Taylor, and points to violence before the rise of the city states as having had particular, local reasons rather than being a universal way of life.

Similarly the picture Mithen describes as regards artefacts is extremely varied. A site now under Lake Assad from 9600-8500 BCE produced clay and stone female figurines, and the skulls and horns of wild cattle were buried under doors and in walls. At the same time, a ritual centre nearby had carvings of animals and strange pictograms, for example of a human arm. There was no sign of a mother goddess, but there was a limestone carving of a male with an erect penis. Was this a fertility symbol, like female fertility figurines? Another nearby site produced stone figurines of birds of prey, and pictograms of animals.

As another example, in Germany from 13,000-11,000 BCE engraved slates were found, of animals and females. There were birds, seals, woolly rhinos and lions. The pictures of women ranged from realistic to almost abstract; some were of single figures, some of groups of three or four, some of ten women suggestive of dancing. In one group there seems to be one carrying a child on her back, her breasts swollen with milk. It is not clear what they were for, and Mithen suggests perhaps they were simply drawn to celebrate women.

The massive temple complex at Gobekli Tepe in Turkey, the oldest known example of monumental architecture, was constructed about 11,600 years ago before the rise of farming. The pillars show bas-reliefs of animals – gazelles, snakes, foxes, scorpions and wild boars – rather than goddesses.

With regard to the "Kurgan hypothesis" of Marija Gimbutas which used to be widely accepted - the idea that the proto-Indo-Europeans came sweeping down out of the Russian steppes about 3000 BCE - Colin Renfrew suggested in 1987 through his studies of archaeology and linguistics that they actually originated in Anatolia (part of Turkey) and/or Western Asia in about 7,500 BCE. Anatolia is the area in which Catalhoyuk was located, whose finds have so often been used to support the idea of the Great Goddess. The Indo-Europeans spread across the whole of Europe and into some parts of central and southern Asia as Neolithic migrant farmers. It was thought that Mesolithic populations survived in some pockets, for example where they speak Basque and Finnish.

This picture has in turn been altered, as DNA studies have shown that language and way of life can spread by adoption as well as conquest. It is now thought that the immigrant farmers are responsible for about 20% of European genes. The question of Indo-European origins is still not entirely settled.

Today it seems that few archaeologists, even feminist archaeologists, agree with the idea of universal goddess worship. In 1998 Lucy Goodison and Christine Morris edited a book of articles by female archaeologists to address this question - *Ancient Goddesses: the Myths and the Evidence*. In their introduction they said that "while the use of the 'Goddess' as a *metaphor* has been inspirational for many, the attempt to reconstruct a *literal* past has appealed to authoritarian attitudes and fundamentalist principles which we find deeply troubling".

For them the archaeological evidence suggests a great variation over time and space in both in social organisation and with regard to artefacts. They also feel that while the figure of the Great Goddess is an important archetype, by concentrating on her it cuts off access to other ways of being female, other archetypes.

Goodison and Morris' book reveals a wide diversity of evidence. Some of it seems to contradict the Great Goddess scenario, and some supports it. It certainly seems to be the case that there have been many places where the Goddess was worshipped, sometimes alone, often among a diversity of deities, and that the social position of women has varied in different societies. At times women have been able to hold a great deal of power and independence, but this is not guaranteed by the presence of powerful female deities. There does seem to be a broad link between the domination of certain incoming peoples, the domination of their supreme male god, and the domination of patriarchy.

How to interpret the most ancient figurines such as the Venus of Willendorf, from the period 40,000 to 5,000 BCE in Europe, is in dispute. Large stomachs and pendulous breasts may not necessarily have meant pregnancy and lactation and therefore fertility. Perhaps the figures denoted womanhood rather than motherhood. What assumptions are we making about gender – for example does fat equal female? In fact most figurines are not easily assigned to male or female gender and some have attributes of both sexes. Many figurines have no sexual characteristics at all, and many are of animals. There is enormous variation in form and decoration.

Some figurines have been found in rubbish pits, which does not seem to suggest religious use, but some are under house foundations, which does. It can be difficult to establish the use of a particular excavated building and therefore the significance of a figurine found there. In Mesolithic Bulgaria, prestige items were found in male graves but no figurines; half the figurines in the village were female but no prestige items were found in female graves. This suggests that even if they were goddess figurines, the status of females, human and divine, was low.

Many different possible uses for ancient figurines have been suggested: toys, dolls, sexual images, markers of special spaces or special times (eg childbirth), people (goddesses, ancestors), ritual equipment, talismen, personal ornaments, educational aids. Modern Western ideas separate the sacred from the profane or secular very sharply, but the lines of separation were likely to have been much more blurred in earlier times. Thus figurines may have had multiple roles. In essence, archaeologists are saying we don't necessarily know why the figurines were made or how they were regarded and we may never know!

In early Crete, in the period 7000-3000 BCE, figurines are mostly found in debris, which gives no proof of their divinity. Most are female but many are male or animal. Later there are many figurines found in graves, which suggests religious significance. There is evidence of ritual but not of personified deities. There are images of plants, animals, the sun, dancing.

The first Cretan palaces appear in 1900 BCE and now images are more anthropomorphic, with fewer images of the natural world. In 1600 BCE snake goddesses appear, deliberately buried in the foundations of the palace. They are associated with crocuses with frescoes of crocus gatherers around goddesses – saffron is used in medicine, cooking and dyes and was evidently being cultivated. There are images of worshippers (mainly women) approaching the deity, or (possibly) having a vision. Later still the former symbols such as snake, axe, plants become part of the god/goddess; there are many deities but most are female. Female divinity has grown with time rather than diminished.

James Mellaart had excavated the ancient settlement of Catalhoyuk in Turkey from 1961-66, and his work was supportive of and inspirational to the idea of a matrifocal, Goddess worshipping society. In the early 1990s one of Mellaart's students, Ian Hodder, began further excavations at the site. What he has found has contradicted many of Mellaart's interpretations. Lynn Maskell (a former student of Ian Hodder) states that Mellaart was making assumptions rooted in his own culture, including the idea of the evolutionary stage of the Great Mother Goddess, and that unknown items too easily get placed in the 'ritual or religion' category. She also feels early archaeologists were too influenced by finds from classical Greece, expecting to find a similar picture of gods and goddesses.

Maskell reports a huge variety of figurines from Catalhoyuk, male, female, sexless and animals, and ones formerly said to be birthing not, in fact, showing clear signs of giving birth or in the birthing position. There are many twin-headed figurines that have been ignored; she also finds evidence of composite beings such as the shape of a bull's muzzle with a vulture head within. She sees no reason to assume that they represent the Goddess. There are broken stalactites which could be phallic objects.

Maskell says that modern excavations can't separate out shrines from non-shrine buildings, so there is no evidence for multiple Goddess temples. There is also no evidence for large scale public construction. Weapons have been found and individuals with head wounds. Grave goods show inequality of wealth and figurines are never found with burials. Thus the picture of the peaceful, goddess-worshipping, non-hierarchical society seems to have been wishful thinking in this case, and by selecting the female figurines the early interpreters of Catalhoyuk rather distorted the evidence.

In summary, the evidence appears mixed. Whether or not societies were peaceful before the rise of city states seems to depend on the circumstances in which they found themselves; with the Urban revolution came warfare. The interpretation of findings to show non-hierarchical societies and equality of women seems suspect now. Physical evidence of goddess worship before writing seems much more ambiguous, and there seems rather a picture of diversity – different societies having different cultures and patterns of worship, which involved male, female, androgenous, animal and abstract images. Greece and Rome show that there could be goddess worship with influential priestesses but still the position of women could be dire.

Whatever the overall picture may be, there are clear examples of goddess-worshipping cultures in which we can trace her gradual defeat by male deities. A detailed look at the evidence from Mesopotamia of the period 3,000-1,000 BCE is possible, with cuneiforms texts available as well as seals and writing on figurines and the figurines themselves. Originally there were mass produced nude female figurines in terracotta which evidently belonged to the common people (could they be personal guardian spirits? good luck charms?);

then there were the many goddesses of the temple or priestly religion; and finally the royal religion.

Deities (and their consort) in Mesopotamia seem to be associated with a particular city, and could be male in one town and female in another. Of 41 Old Akkadian temple hymns, 16 were to goddesses. There was a Vizier deity Ninshubur who was female when serving female goddesses, and male when serving male gods. It has been suggested that originally the deity had no gender; the word for deity and for lord/lady had no gender.

In the early days of the Sumerian civilization there was a large variety of goddesses – associated with birth, the birthing hut, wet nursing, grain cultivation, ewes, agriculture generally, the sea, beer, wine, weaving, pottery, jewellery, metallurgy, healing, dreams, punishment, weights and measures and lots, lots more. Pestilence and destruction had male gods. The heavens were generally male, including the moon, but the goddess Inanna (nin-an-na – Lady of Heaven) was associated with Venus, the morning/evening star. The earth was generally female and the realm of the dead belonged to Ereshkigal, sister of Inanna. Inanna of Uruk in the oldest inscription was the giver of the supreme rule of Sumer. She was later replaced by Enlil (male) of Nippur, another city. Rulers were referred to as the spouse of Inanna, and other goddesses were protectors of kingship, patronesses of royal families. There is evidence of the sacred marriage ritual between king and goddess, or priest/priestess with deity. Thus kingship was renewed and prosperity assured.

A text from the mid-third millennium talks of the "lady of all the me's (divine principles and powers), resplendent light, righteous woman clothed in radiance, beloved of Heaven and the Underworld'. However Inanna was supposed to have stolen the me's from the god Enki, god of sweet water. This may reflect the rise to power of one city over another.

In the case of ancient Sumer, at one time it was thought that Proto-Indo-European tribes had invaded as early as 3200 BC, thus arriving just after the invention of writing. This would mean that pretty well all of recorded history was written from the Indo-European, patriarchal point of view – the invaders from the north imposing their kingship and leading male gods. However, it now seems that the archaeological record shows continuity from peoples already living in the region (the Ubaid culture) which dates back to about 5,000 BCE.

Thus any early imposition of male gods cannot be explained through Gimbutas' Kurgan (northern invasion) hypothesis.

The Early Dynastic period of Sumer (from about 3,000 BCE to about 2334 BCE) saw the flowering of the city states and the spread of Sumerian influence and culture beyond its borders, probably largely through trade. There were both linguistic Sumerians and linguistic Semitic people, for example the city of Kish which was to the north of Sumer was in the Semitic area called Akkad. The city states vied for prominence (and also had to fight off occasional invasions from foreigners eager to get their hands on Sumerian wealth), and in particular Uruk, Ur and Kish all produced kings of Sumer and Akkad. Nippur on the other hand was the religious capital and the city of Enlil, Sumer's supreme god: "chosen by Enlil of Nippur" became one of the king's titles.

It seems that although there were linguistic differences between Sumer and Akkad, the culture of the two areas was one. The Semitic peoples are now thought to have been there from prehistoric times, as were the Sumerians, and once again this is not a case that fits the Kurgan hypothesis.

In 2334 BCE the Semitic king Sargon became king of the Akkadian kingdom of Sumer and Akkad. The Akkadians imposed their Semitic goddess Ashtar (Ishtar) who was merged with Inanna, perhaps to make her more acceptable to the Sumerians. There were originally two universal deities, male and female, but later a host of deities developed, with a specific personal god plus female guardian deity and male guardian deity (later demons). War and victory were mainly female goddesses. Ashtar was Lady of Battle, love goddess and motherly protector.

Sargon is quoted as saying in one saga that he became king after "Ishtar granted me her love"; but he also called himself "anointed priest of Anu" and the "great *ensi* of Enlil" (both male gods). His daughter Enheduanna was made priestess of the male moon-god Nanna, not the goddess Inanna, although she is known for her hymns to Inanna and called on Inanna for help.

In the second millennium BC there was a marked marginalisation in the role of female goddesses. Many became male. In Babylon its god Marduk became the supreme god, and 'ishtars' became a generic name for goddesses. In myth the male god ploughed the goddess earth, symbolising the subjection of the female. However also in myth, the couple produce their son, the cattle god. The earth desires the son and kills the father to marry the son. Eventually even fertility acquired male gods – Baal for the fertility of crops and El for the fertility of humans.

It is hard to draw firm conclusions from times before the invention of writing, which occurred about 3,300 BCE in Sumer (about 100 years before it was invented in Egypt). However, it seems that by the time of the city states, male gods were perhaps in ascendancy but both male and female deities made up the Sumerian pantheon, and each city had its own particular protecting deity – which might be male or female, but more often male. Female goddesses were still important, particularly by way of the ancient rites of the sacred marriage which validated the king's reign and ensured the fertility of the land.

As time went on, male gods became relatively more important. This seems to mirror the growth in size of the state, from city to country to empire, and the growing importance of conquest. With the establishment of the Babylonian empire in about 2000 BCE the victory of the god Marduk over Tiamat, goddess of chaos in the creation myth, became more important than the ritual of the sacred marriage; the former was ritually celebrated every year and the latter discontinued. Here, clearly, the defeat of Tiamat can be seen as the defeat of the Great Goddess. We shall see in the next chapter how the Goddess was defeated in another area of great significance to modern Western women – Israel, land of the Bible.

CHAPTER 12 – Defeat of the Great Goddess

It seems likely that a major change in religious ideology accompanied both the Neolithic Revolution with the rise of farming and the Urban Revolution with the rise of the city states.

The arrival of the Neolithic way of life with settled agriculture about 10,000 years ago transformed society. There was a new concept of private ownership of the land, and men could now gain power by laying claim to the surplus produced by agriculture. These new ways of being meant new attitudes to the natural world and within society. As we saw earlier, generally the status of women declined.

With the Urban revolution from about 5000 years ago there was a transfer of populations to towns and cities and the distancing of a ruling class not only from nature but to a large extent from farming. Great stores of wealth were amassed, and huge palaces and temples and monuments to the rulers were built. There was a tightly controlled bureaucratic society where most lived at subsistence level. There were wars between city states which grew to encompass countries and then empires. In places where the Goddess was worshipped, this seems to have often led to a fall in her status.

There were movements of Indo-European peoples and their influence on religion can be seen clearly in places such as Greece and India, but this cannot be used as a general explanation for the waning of the Goddess. It seems that it was more often the developments of civilisation that accompanied the fall in status of the Goddess, rather than conquest by different peoples.

Most archaeologists seem to agree that there was a change in how the world was viewed, accompanying the change both to a sedentary lifestyle and to a dependence on farming. Humans must always have been aware of both benign and destructive aspects of nature, but for most of our history, nature was the storehouse from which we gathered whatever we needed for life. After the Neolithic Revolution, there was a difference between Nature, Out There and untamed, still a source of danger but increasingly not the source of food; and Inside (house or village or farm), the domesticated area, probably marked off with fences and walls and boundary markers, where there was safety and food.

In the Paleolithic (Old Stone Age) plants and animals reproduced without any intervention, and humans then gathered what they needed. There may well have been the belief, as we have seen in existing hunter-gatherer tribes, that religious rituals of some sort would help to ensure ongoing fertility of the land or the return of the rains, and the enigmatic figurines from Paleolithic times may have been to do with this; but it was not about direct human intervention.

Once the Neolithic Revolution had occurred, and there was no going back, direct human intervention was essential. Crops would not grow unless planted, grain could not be gathered without reaping and threshing. Herd animals could not be kept without constant care. Fertility had become something to be worked on and controlled. Nature had become something to be kept out, in the form of weeds among the crops and predators coming after the herds.

Further control was needed in the form of private property. If you depended on the produce of a particular field, and had spent all your labour on it, you wanted to ensure that no-one else wandered in and started gathering the crop. People and social relations had to be controlled too.

The line between order and chaos became very important. Chaos was Out There while Order was In Here, and Chaos had to be kept Out There. No wonder the creation myths were now about the first gods or heroes slaying the personification of Chaos.

The Babylonian creation myth is written down in a long poem called *Enuma elish* from its opening sentence, "When on high…" In it, the hero-god Marduk slays the goddess Tiamat who represents Chaos, and from her mutilated body he creates the earth and the sky. Marduk then proceeds to set the universe in order, placing the gods in their dwelling places and creating humans to serve them. The poem was originally composed about 4,000 years ago, though the age of the myth is unknown. The *Enuma elish* was recited by the priests of Babylon on the fourth day of the New Year Festival (in April) for perhaps nearly 2,000 years, to ensure that the forces of Chaos did not even now challenge the Order of the gods.

The Fertile Crescent, which produced, in ancient Iraq, the kingdoms of Sumeria, Akkad and Babylon, was fertile because of the waters of the Tigris and the Euphrates. They brought silt down from the northern hills and periodic flooding spread the rich silt over the floodplains. The water, through irrigation channels, could be diverted through fields to give a steady supply to the crops and for herded animals. The Neolithic Revolution had brought food surpluses as never seen before, with the development of high-yielding strains of plant and animal. However, the possibility of disaster was never far away.

Weather cannot be depended on, and there were times of drought and flood. Plant pests proliferate where a single crop is grown, and disease spreads among close packed herds. Irrigating the land meant a slow accumulation of salts that gradually made the land infertile and unusable. If things go wrong for a hunter-gatherer tribe in one area, they can move to another. When you have hundreds of species to choose from for your food, the loss of one or two probably doesn't really matter. But when you are a farmer, if things go wrong it can mean complete disaster, starvation and death. Even in the West, it is not so long ago that the people would starve if there was a crop failure, and they were not rare.

Georges Roux suggests that constantly living on the edge of disaster explains the pessimistic nature of the Sumerian people's religion. They weren't put here to enjoy the beauty of the earth, but to serve the gods (or goddesses); if things went wrong it was a punishment from the gods for infringing some petty rule of theirs. Humans could only beg for mercy and hope to propitiate the deities through offerings and sacrifices and strict observance of the rules.

This longing for order rather than chaos, a predictable world where disaster would not strike, may well have affected the status of women. Women were originally associated with nature, because of their monthly cycles (so close to the moon's), their bleeding, their involvement in the mystery of birth and thus, by association, with the mystery of death (people buried in the foetal position as if for rebirth in the womb of the earth), also their knowledge of and association with the fruits of nature – plants and herbs for food, medicine and crude contraception. This association would originally have been a source of respect.

Once nature became a dangerous beast to be tamed, then woman's association with nature would have made her suspect. Today there are many qualities which are traditionally gendered – women are said to be intuitive, nurturing, feeling, multi-tasking, right-brained; men are logical, thinking, judging, single-minded, left-brained. Riane Eisler saw this as symbolised by the blade – the male who takes away, dominates; and the chalice – the female who nurtures and gives.

Of course these are stereotypes, but perhaps based on the roles arising from our biology. Although there will always have been a crossing of gender boundaries, on the whole women will have needed to be empathic, tolerant, intuitive and so on in order to be good carers of children, whereas because of their greater strength, men have tended to be the ones who must be brave and prepared to kill to provide meat and protect their kin. As nature became more suspect, at the same time as men became more socially powerful, perhaps the qualities traditionally associated with women became less and less valued.

There are certain images found at Catalhoyuk from perhaps 7,000 BC that archaeologists who have seen them, tend to find disturbing. This is the ancient town that Gimbutas used as supporting evidence for the existence of a matrifocal, goddess-worshipping, early Neolithic society. Some of the houses there have horned bulls' heads protruding from the inner walls. Although this has been claimed as an aspect of Goddess worship – or at least complementary to the Goddess, as Eisler suggested – we can't really know why they are there. The disturbing image is of paired (and sometimes single) mounds of clay also found on the inner walls of the houses, sometimes thought to represent breasts, with vulture beaks appearing where the nipples should be, or occasionally foxes' teeth, wild boar tusks or a weasel skull. Mithen describes this as "motherhood itself violently defiled".

In fact there may be no good reason to see the mounds as breasts. Mithen himself muses that the vulture beaks may not be meant to be sinister, since birds of prey had been revered at other sites. However, he feels that the people of Catalhoyuk gave the impression of fearing and despising the wild. Timothy Taylor says the sculptures are "creepy and suggest a community ill at ease with the world". As always, we cannot know for sure what was intended by the original creators of these images. However, if vulture beaks are protruding from women's breasts, this may be an early indication of humans' alienation from the natural world around them. Rather than this early Neolithic town being a place of matrifocal harmony, it may show the beginnings of a human/nature split which accompanied and contributed to women's oppression.

The Bible is the set of religious writings that most affects women in the West. As with any culture's collection of myths and histories, we can trace a story of changes and conflicts between different traditions. Goddesses are present in the Bible to a surprising extent. The Bible is often treated as a straightforward historical document but this is not so, particularly since the concept of history was understood differently in earlier times. Modern scholarship has much to say about how the book, or series of books, evolved.

The Bible in its final form starts with stories of the creation and of legendary times from Adam and Eve to Noah and his descendents. Then come the stories of the Patriarchs, Abraham and his descendents, Isaac, Jacob (later renamed Israel) and Joseph, and how their people went down in to Egypt and were enslaved. This all happened between 1800 and 1200 BCE. Next Moses liberated the Israelites from slavery and brought them back to the Promised Land, or Canaan. The Israelites conquered Canaan and took over the land, destroying the old religion.

In about 1000 BCE David established a monarchy at Jerusalem in the south, uniting the country by conquest. He was followed by Solomon; on his death, the kingdom was divided into a northern kingdom, Israel, and a southern kingdom, Judah. In about 850 BCE King Ahab of Israel died. This was the time of the prophet Elijah. Further prophets (Amos and Hosea in Israel and Isaiah and Micah in Judah) followed from about 760-690 BCE. In 722 BCE the northern kingdom fell to Assyria and the people of the northern ten tribes were carried off into exile, to be forcibly assimilated and lost to history. At this point the religious traditions of the north were brought to the south.

In 587 BCE Jerusalem fell to the Babylonian empire, and many Jews were carried off into exile there, until Cyrus of the Persian empire conquered Babylon in 539 BCE. Many of the exiled Jews returned and a second Temple was built. In 332 BCE the Persians were defeated by the Greeks, led by Alexander, and Greek domination began until (after a brief interlude) the Roman empire took over. In 70 CE the Romans destroyed Jerusalem and the Temple.

Early modern scholars found that there are repetitions of stories in the Bible – if you read Genesis for example, you will find there are two, conflicting stories of the creation. They found that one set of stories refers to God with the sacred Hebrew name YHWH – Yahweh or Jehovah. The other uses a more generic term, Elohim meaning "the divine". These two different traditions were called J and E, with J in the south and E in the north. After the fall of the northern kingdom E was brought south by refugees.

Further double sets of writings were noted, one set being concerned with priestly matters of worship, purity, ritual, law. This was called P. A final distinct tradition, in which God insists on being the only deity and punishes those not faithful to him alone, was called D (after the Book of Deuteronomy).

Abraham was an Amorite, linguistically a North Western Semitic race, their languages including Ugarite (from Ugarit in Syria), Aramaic (the language Jesus spoke) and Hebrew. The Amorites were a patriarchal, patrilineal race, of the Indo-European type we met previously. They had sky gods with an emphasis on lightning, storms, mountains.

Some of the original mythology of Canaan is available to us from the archaelogical finds at Ugarit, an ancient city on the Syrian coast. In one story from about 1400 BCE Baal, the god of storms and fertility, and Yam, the god of seas and rivers (representing chaos), lived with the High God El. At the Council of El, Yam demands that Baal be delivered up to him. Baal defeats Yam and is about to kill him when Asherah, El's wife and mother of the gods, intercedes on his behalf.

In another myth Baal slays the dragon Lotan, Leviathan in Hebrew, which also symbolises Chaos and perhaps the female aspect. Baal dies and descends to the underworld, domain of Mot, the god of death and sterility. Baal's lover and sister, Anat, goes in search of him; when she finds his dead body, she holds a funeral feast for him. She kills Mot, burns and grinds him like corn and sows him in the ground. Later Baal is brought back to life and restored to Anat.

Here we have the common myth of the dying god, lover/brother of the female Goddess, who dies and is reborn in the round of the seasons. As elsewhere, this was celebrated by ritual sex in ancient Canaan. Anat, a maiden goddess, was a passionate lover, a warrior and huntress, the patroness of warriors and kings. As well as Queen of Heaven she was called Mistress of all Gods. Under the Egyptian empire in 1300-1200 BCE her worship spread outwards.

Karen Armstrong in *A History of God* thinks that the early patriarchs of the Bible were not monotheists, believing in one God only, but would have believed in the existence of many gods; certainly the deities of Canaan such as Baal and Anat, and more distant ones such as Marduk of Babylon. Abraham's god was probably El, since he introduced himself to Abraham as El Shaddai (El of the Mountain), which was one of his traditional titles. El was the chief god of Canaan and his name is preserved in many names in the Bible, including Israel. He was a comparatively accessible and human-like god, like other pagan gods of the time.

There are suggestions in the Bible that it was Moses who introduced Yahweh to the Israelites, Yahweh having appeared to Moses in the burning bush. Yahweh was a warrior god (Yahweh of the Armies). He started off quite a cruel and violent god, as so many powerful gods (and goddesses) were in those times, though he later became more a god of social justice and compassion and ultimately of transcendence. Moses declared that Yahweh was the supreme god. This was the southern tradition – the J strand of the Bible.

The Hebrews were clearly worshippers of the old Amorite gods and goddesses. There are many references in the Bible to Baal (a fertility god) and El (whose symbol was the bull). King Solomon had a large bronze basin in his temple to represent Yam, god of the sea. It seems likely that the early Hebrews were polytheistic, with a chief, male leader god which probably reflected the male leaders of the tribes. However, goddesses were still worshipped too.

The asherahs referred to in the Bible as being associated with the worship at the altars of idols, were the sacred trees or wooden pillars of the worship of the goddess Asherah. The sacred tree might have been a sycamore fig or black mulberry, which has reddish fruit in clumps. This might be the source of the Tree of the knowledge of good and evil. The snake was also a common goddess symbol, for example Ishtar had a serpent-entwined staff. This may be for several reasons – the fact that snakes shed their skin and are in a sense reborn; and perhaps snake bites were used as a kind of hallucinogen for religious visions. Thus the story of Adam and Eve may include a denigration of the old ways.

In 622 BCE Josiah was king of the southern kingdom of Judah. The two previous kings had allowed the people to worship the old gods alongside Yahweh, and one of them had put an effigy to Asherah in the Temple. Asherah was seen as Yahweh's wife, as she had previously been El's wife. Now Josiah, a Yahwist, determined to put the Temple to rights. During building work it was claimed that a senior priest had found an old manuscript which was supposed to be Moses' last sermon. This was probably the inspiration for "D", the third main source of the Bible. This included a version of the covenant in which Yahweh demanded complete loyalty and rejection of all the other gods. Psalm 82 has Yahweh in the Council of El, saying he will kill the other gods.

Now, with D, the reformers rewrote history. The Bible says that after the Exodus from Egypt, the Israelites were told by Yahweh to conquer the Promised Land (Canaan) and to show no pity to the inhabitants. They were to wipe out the Canaanite religion, killing the people and destroying their religion. Everybody and everything was to be killed except the sheep and the asses and the virgins. The Israelites were all to allow their own captive virgin to mourn for her murdered parents for one month, then they could have intercourse with her and take her as a wife.

Why this emphasis on virgins? The Israelites were a patrilineal society; they measured descent through the male, not the female. Thus it was of the utmost importance to control who exactly the father was. The only way to be sure of paternity was to make sure that a woman could not have sex with anyone else, which in turn meant that virginity before marriage and loss of freedom after marriage was essential.

In fact archaeological evidence increasingly suggests that this conquest of Canaan never happened. Archaeologists Israel Finkelstein and Neil Asher Silberman in their book *The Bible Unearthed* suggest that there was no Exodus to Egypt and return to the Promised Land, but that the Hebrews had always lived in Canaan and never left it.

They maintain that the archaeological evidence shows the gradual evolution of separate Semitic societies in the lands of Judah (to the south) and Israel (to the north). Both were marginal, mainly hilly countries – though Israel had more fertile land than Judah. The story of the Exodus may be based on the fact that Semitic people did sometimes move down to the fertile lands of Egypt when the crops failed in their less dependable homeland.

Finkelstein and Silberman believe that the key event was the collapse of the northern kingdom, Israel, after its defeat by the Assyrians in 722 BCE, leaving something of a power vacuum there. It would have been in the interests of Josiah, king of Judah, to maintain that Israel and Judah should really be one kingdom. Thus the story of the violent conquest of the whole area, and its further unification under the southern king David, promoted Judah's claim on the northern kingdom. The convenient "discovery" of the document D in the temple would alter history and legitimise Josiah's claims.

Whatever the exact truth of the matter, King Josiah was now determined to impose a new state religion of one supreme God, Yahweh. He had anything linked to the old religion taken out of the Temple and burned. There had been earlier references to sexual customs, eunuch priests, and grieving for Tammuz or Baal as the dying consort. However, the priestesses of the temple, with their sacred sexual acts, were now denigrated as temple prostitutes. The prophet Hosea calls his wife Gomer a prostitute; she was priestess of the temple of a goddess. The apartments of the priestesses were burned. All the ancient shrines in the country were destroyed. From now on, priests could only sacrifice to Yahweh in the Temple in Jerusalem. In the book of Deuteronomy (Chapter 13) the people are told that if their brother or son or daughter or wife or friend suggest serving other gods, they must kill them.

The fierceness and strength of the new patriarchy demanded the control of women's sexuality to an even greater degree. The laws of the book of Leviticus and Deuteronomy show how the new sexual rules and morals were imposed to protect patriliny. See for example Deuteronomy Chapter 22: If a woman committed adultery, she and her lover were to be put to death. If a husband says he can find no sign of virginity in a new wife, and nor can her parents, she is to be stoned to death by the men of the town in front of her father's house "to purge the evil from among you"(Deut 22:13-21). If the daughter of a priest "plays the whore" she must be burned to death (Leviticus 21:19). Menstrual blood is now regarded as powerfully bad rather than awesome – "do not approach a woman to have sexual relations during the uncleanness of her monthly period" (Lev 18:19).

Lilith was originally a goddess who was described in a Sumerian tablet as "the hand of Inanna" – she went out and brought men into the temple. Now in some Hebrew literature she is described as the first wife of Adam, who refused to lie beneath him and obey his command. In the Kabbalah she is transmuted into a female devil, who provokes ejaculation from which she makes demons and illegitimate children. When the man dies, the latter go after his inheritance – a fear resulting from patriliny.

Women become second class citizens and their sexual nature is tainted with fear and disgust. To further control them, they are the property of their father and then of their husband or other male relative. In ancient Sumer, if a man took a second wife after the first had borne children, he was to be put out of the house without any possessions. Now a man had only to write a note to divorce his wife, and she owns nothing.

We see the beginnings of the emphasis on the intellect, on language and logic, as opposed to feeling and intuition. There is more and more emphasis on the Word (even in the still, small voice). Naming things was seen as having power over them; thus in one of the creation stories, Adam named the animals and had power over them. When Moses asked Yahweh's name he replied "I am that I am", which some translate as Hebrew for "Never you mind!", or, it is not for you to name Me. Naming things divides them up, categorises them, controls them and brings them under the force of logic; it tames nature. Included in the control of nature was the control of women.

This is not to suggest that there were no gains. We now see in the Bible the beginnings of calls for social justice and compassion (the rise of individual conscience). Polytheistic worship had tended to be associated with a certain fatalism, and gods and goddesses alike were praised for all their deeds whether they favoured the human fate or harmed it. The gods were seen as capricious creatures who had to be appeased to try to avoid disaster, for disaster appeared often. In the hymn of the priestess Enheduanna to her favourite deity, Inanna, she says: "Vicious dragon you spew, venom poisons the land, like a storm god you howl, grain wilts on the ground, swollen flood rushing." (Meador, 2000). Yet Inanna is "cherished in heaven and earth". Now arose the idea that individuals had a religious duty to promote the welfare of those less fortunate.

In the end the southern kingdom could not stand. The new Babylonian empire invaded Judah and carried off the king and others of high class to exile. (Some who fled to Egypt blamed the new religion for the defeat, and returned to baking cakes for the Queen of Heaven.) The Jews in Babylon were not forced to assimilate, and re-established Yahweh and his temple when they were later returned to Jerusalem when Babylon was defeated.

When Cyrus, king of the new Persian empire, allowed the Babylonian exiles to return home, they took their more developed Judaism with them and imposed it on those who had remained behind. The source "P" was written after the exile and again rewrote history. A more sophisticated version of Yahweh was added: you can no longer see Him, only the "glory" of His presence. It is P that provides the second, longer story of creation in Genesis. As in Babylonian myth, Yahweh makes the world out of the formless and chaotic waters, but he doesn't have to fight for supremacy as Marduk did, or make the earth out of the body of the goddess Tiamat. Yahweh is in a sense outside and beyond his creation, which is completely at his disposal.

Yahweh was established as the one and only God; a transcendent god rather than a pagan nature or warrior god. Whereas pagan gods tended to be much like ordinary humans, but with supernatural powers, Yahweh was now seen as something beyond human comprehension. And with the establishment of the one, true, male God, the female element in the divine was largely lost.

The suppression of Goddess worship was completed by the might of the Roman Empire after it was converted to Christianity by the Emperor Constantine. The worship of Asherah was now illegal by Roman law. In 380 CE Emperor Theodosius closed the temples of the goddess at Eleusis (where the Eleusian mysteries took place – they were based on the myth of the abduction of Persephone), also at Rome (Isis was worshipped there at this time) and at Ephesus (Diana/Artemis). In seventh century Arabia Mohammed ended the worship of the Goddess (Al Lat and Al Uzza).

When the Great Goddess is missing, there does seem to be a Goddess-shaped hole left behind. Even though the Proto-Indo-European God eventually took over large parts of the world, with (in Christianity) his male son and male Spirit and mostly male prophets and heroes, there were still female saints and most of all, the Virgin Mary. Of course she is on the whole a very poor role model for women, one of her outstanding features being her essential passivity and asexuality. But it is interesting that as time went on she became more and more important in the Catholic Church and ended up as the Queen of Heaven – in the Sumerian language, Nin-an-na, or Inanna.

CHAPTER 13 - Reclaiming the Goddess: Maiden, Mother, Crone

Millennia of subjugation must have left its mark on most of us. We are not helped by the negative role models we encounter, for example those Biblical figures introduced to us by our families, our school or just absorbed through our culture – the passivity of Mary, the sinfulness of Eve, the harlotry (actually not present in the Bible) of Mary Magdalene. Although modern readings of these women can be quite different, the commonly encountered interpretations give a negative image of womanhood. That is one reason why we need to look back to our charter myth – before the male gods conquered and subdued the Goddess. We need a balance in the West to the Christian tradition and to the Jewish and Islamic traditions that share the same roots.

We need symbol and metaphor which can bypass our conscious mind, and go straight to the unconscious with archetypal images of power and potential that are deeply female. We don't have to take any of this literally. But by examining and identifying with the aspects of the Great Goddess that we can find in our past, by imaging and visualising her before us, in many different roles, we can understand and reclaim deep aspects of ourselves – giving us the strength to face the journey of ageing with courage.

We also need to reclaim the reconciliation of opposites that we find in goddess worship. In the Biblical creation myth written after the exile in Babylon, when Yahweh was supreme, God separated night from day, water from dry land, light from darkness. The monotheistic male religions tend to divide the world up into opposing qualities: good and bad, light and dark, life and death, success and failure, doing and being, male and female, and they make the one to be desirable and the other to be rejected and feared. Since we can't cut the world in half and get rid of the bits we don't like, this only sets up conflict within ourselves.

Goddess worship, however, implied that all was accepted under her wing. She gave birth to us but also received us back in death; we walked in the light but she also ruled in the underworld; her yearly ritual marriage and the death of her consort with rebirth in the spring played out the ever rolling cycle of the seasons with growth, harvest, dying back, and growth again.

Goddess worship embraced and respected the natural world as well as cultivation and agriculture. In early societies people were more in touch with the endless rhythms of life – from daylight to the darkness of night (with fire the only artificial light); from the abundance of summer and the fruits of autumn to the scarcities of winter; from years of health and abundance to years of starvation and disease. People were more aware of birth and death and the seeming arbitrariness of natural disasters.

We can assign the unwanted half of the opposites of life to the devil, but that won't get us very far. Or we can strive to accept all that life throws at us, even if we still have our preferences, and see it all as part of creation in the person of the goddess. What we see as the opposites may even be personified in different deities, for example Inanna as goddess of life, Ereshkigal as the goddess of the underworld, but one was not necessarily seen as "better" than the other. The darker goddess was not seen as evil but as part of the pattern of life. In India, Kali the destroyer goddess is also worshipped as a mother goddess.

Embracing the opposites would help us not only to find the courage to strive and fight for and do – think of all the terrifying warrior goddesses – but also to embrace, when appropriate, the traditional female qualities of holding still, pregnant with possibility but waiting until the time is ripe for movement. We would celebrate being, with no outward effort, as much as our present society worships fame and success. We would honour our skills in nurturing, perhaps those younger or weaker than ourselves but also our sisters and friends and the earth.

We could also do with blurring the separation between sacred and profane that is found today. One reason why Celtic Christianity is so popular now is because their prayers quite naturally brought the sacred to everyday tasks such as milking the cows, and covered every hour of the day. It wasn't a question of going to church on a Sunday morning and forgetting about the sacred for the rest of the week. Remembering the sacred in all our tasks brings a lightness, a sense of support and courage, a kind of magic to our life.

We can find diversity in the aspects of the Goddess and choose the one that seems to fit us now, today, in whatever particular situation, mood or need we find ourselves in. In particular we want images of ageing women. While the goddess can be seen as Creator/Destroyer, another common division is that of the Triple Goddess – representing the three aspects of the moon: waxing (maiden), full (mother), and waning (crone).

The time the moon takes to go through its phases is about the same as the duration of a woman's menstrual cycle. This may be one reason for the association with the goddess. The waxing moon (whose image in the night sky one can cup in the right hand) is associated with the youthful goddess, sexually active, independent, perhaps a warrior and/or huntress, also representing new beginnings (Persephone, Brigit). The full moon is associated with the Mother, representing fertility, fulfilment, productiveness, nurturing and protection (Demeter, Lakshmi). The waning moon (whose image one can cup in the left hand) represents the older woman, wisdom, divination, the underworld (Hecate, Hel). Robert Graves (1948) referred to them in this way: "the New Moon is the white goddess of birth and growth; the Full Moon, the red goddess of love and battle; the Old Moon, the black goddess of death and divination."

You could say that even today we have a triple-aspected moon goddess in the Virgin Mary: the Maiden when she accepts her role ("Let it be unto me according to thy will"); the Mother when she bears Jesus and later stands at the foot of his cross; and the powerful Crone when she finally takes her place as the Queen of Heaven and receives prayer and petition.

In the following section we will look in detail at a selection of goddesses, and try some visualisations in which we can meet and get to know her better. I have chosen those goddesses which I think are particularly relevant to the crone stage for various reasons, even if they are not crone goddesses. Since goddesses are associated with myths, we may meet archetypes of both goddesses and heroic journeys.

The Triple Goddess: Maiden, Mother, Crone

The form that the Divine takes in the human mind is very much influenced by the society in which we live. In fact, how could it be otherwise? We only know what we experience around us, most influentially as we grow up and then what we are told by the people around us, be that the tribal lore of early times or the newspapers and television of today. What we know, we then project out onto any image of the Divine. Rather than us being made in God's image, God is necessarily made in our image. As the form that society takes alters over space and time, so will the conception of the Divine, including Goddess worship.

Thus the details of a goddess will vary from tribe to tribe, city state to city state and later country to country. She will vary most strongly over time, from hunter-gatherer society to the first settlers to the Neolithic revolution and the development of farming; from tiny villages to towns to cities, from city states to countries to empires. As human focus moved from feeling part of nature, to controlling nature but still subject to its whims in the form of failing crops and natural disasters, to being controlled by civilisation and its complex demands, to having religion hived off as the preserve of a wealthy priestly caste, ideas about the divine must have changed in as revolutionary a way as the society that produced them.

Another complication is that certain archetypes of Goddess appear again and again, just as certain themes appear again and again in different myths. It is not clear whether this is simply because of cultural influences spreading out from the original seat of a particular deity (the worship of Isis arriving in Rome because of the busy traffic of produce from Egypt to Rome), or because there are universal human archetypes and universal conditions that are basic to human experience and thus to myth. An example of this would be myths that explain or symbolise the turn of the seasons, and in particular the loss of fertility and abundance as winter comes on and everything starts to die; a situation no doubt associated with deep-seated fears as to whether spring would ever come again.

Today we look back at the Goddess from our own set of assumptions and needs, and it is impossible to see her clearly. She is still worshipped today, certainly in non-Western societies but also in the West - particularly in the religion of Wicca, but also in a multitude of different sects. The modern overlay influences how we interpret the past, on top of the complication of a goddess having kept her name but changed her nature as her worship moved through time or space. With these caveats in mind, we shall start by looking at the goddess archetypes of maiden, mother and crone.

The Maiden

This is the young unmarried goddess. She is often referred to as a virgin today, but this is a distortion brought about by our modern patriarchal monotheism which celebrates virginity because it denigrates women's sexuality. Think of the various female saints who as virgins chose death rather than to be ravished, and were celebrated for it. Mary, mother of Jesus, is of course the most famous virgin maiden; she even managed to have a child without losing her virginity.

On the contrary, the original maiden goddess was often very active sexually, like Anat, the Canaanite warrior goddess. She gloried in her sexuality, if she did not simply take it for granted. She was young, beautiful and irresistible, and saw no reason (her worshippers saw no reason) why she should not enjoy the pleasures that her body was built for. She was often also associated with fertility, because of her youthful sexuality, and because of the prime importance of the return of fertility each year to a community dependent on crops. Because of her energy, she was also often a warrior goddess, skilled in the hunt and the use of the bow. Thus the maiden represents Springtime and the new blossoming of flowers and leaves, fresh beginnings, renewed hopes, new energy. She might be turned to for starting new projects and finding renewed energy.

Examples of maiden goddesses: Inanna (Sumer), Anat (Canaan), Brigit, Rhiannon, Blodeuwedd (Celts), Persephone, Artemis, Athena (Greece).

The Mother

The Goddess in her most powerful aspect has probably most often been worshipped as a mother. From earliest times, the earth has been seen as the mother who nourishes us, as well as the terrible mother who takes us back into her womb. As we have seen, for some time the Great Goddess was seen by archaeologists only in her mother aspect.

As mother, the goddess is also associated with fertility – often represented as pregnant (though as we have seen, the evidence from the figurines may be contentious), rounded in form with large breasts and buttocks, stomach and thighs. This is a woman of ripeness, maturity, proved fertility. This is a goddess of summer as the fruits ripen and the corn becomes ready for harvesting, and abundance is again assured.

As a mature woman, the goddess is also associated with power. She has the power of her own wisdom and stability; she has a consort rather than many lovers and rules rather than wanders and hunts. She has the power of the archetype of the mother, who reigns over her babies and infants absolutely. The children will often love her passionately and sometimes resent her bitterly, but there is nothing they can do to sway her. We all carry the memory of that initial helplessness and dependence.

The mother loves her family, her husband/consort and her children. The mother is fierce in the defence of her children, ready to give up her life to them. She can be utterly ruthless when necessary, and she will kill to protect herself and those dear to her. She is the goddess that we should call on her help in times of need, when strength and staying power are called for.

The mother is also ruthless in carrying out the laws of the universe. When it is time for death, she will not protect us, but instead welcome us back into the womb of her earth. Only in that way can she give us rebirth.

Examples of mother goddesses: Isis (Egypt), Demeter, Hera (Greece), Frigg, Freya (Scandinavia), Mahadevi (India), Asherah (Canaan), Danu (Celts).

The Crone

When the goddess is linked to the phases of the moon, the maiden represents the waxing moon, the mother the full moon, and the crone the waning moon. She is the mysterious goddess, in some ways the goddess of decline – our decline, but because of her association with death, she is a powerful goddess of the unknown, mystery and occult knowledge.

The crone goddess is associated with the end of the year, the dying off of the vegetation. In Scotland she was associated with Cailleach, the old woman, the north wind, particularly at the festival of Samhain on what is now Hallowe'en. This marked the end of summer, the beginning of winter, the time when the livestock that could not be kept fed throughout the winter were slaughtered.

She is associated with the thin place between life and death, and through this, she was associated with the power over the passage between the two; thus with magic, healing, divination. She was associated with any liminal place, the boundary where you pass from one state to another, and thus with midwifery. She is often seen as a terrible old hag, perhaps malicious, but also as holding the wisdom and personal power that can come with old age.

The crone, while most relevant to us as ageing women, is the most problematic of the three archetypes in historical terms. It is not clear how common she was as a goddess, nor even whether the goddesses were commonly associated with the waning phase of the moon. Robert Graves' description of the three aspects of the Goddess in *The White Goddess* is thought to have been based on unreliable sources, and scholars think he fitted certain goddesses to preconceived ideas.

Many triple goddesses were not maiden, mother, crone, but (for example) three sisters (eg Brigid). It is easy to assume that because a goddess is described as a fearsome old hag, she must be meant to be a crone, but she often turns out to be a mother goddess as well – for example, Kali (India) or the Morrigan (Celts). It may also be that the lines between maiden and mother goddesses was not so clear cut to their original worshippers.

Given that the Goddess is the divine aspect of the female principle, it should not be surprising that one goddess could contain several different aspects of woman according to the circumstances portrayed in the myth. This is particularly relevant to us, since as ageing women we carry within us the three archetypes – the maiden that we once were; the mother that we may well still be (the mother archetype if not an actual mother); as well as the crone that we are becoming.

Bearing all this in mind, examples of the crone goddess might include Hecate (Greece), Ereshkigal (Sumer), Hel (Scandinavia), Baba Yaga (Russia), Cailliach Bheur (Scotland).

CHAPTER 14 - Goddesses for the Ageing Woman: Hera, Hecate, Ereshkigal

This is a selection of goddesses which can be imaged specifically for the older woman, which we shall explore in some detail and then see how we might make use of these archetypes. First there is Hera, who was originally worshipped in the triple aspect of maiden, mother and crone within the one goddess. This represents how we have within us all the stages we have been through, as well as the stages still to come.

Next is Hecate, for a goddess of thresholds and passages – the passage through the menopause; later the passage from a still vigorous late middle age to old age; later still, the passage to death.

Then there is Ereshkigal, goddess of the underworld, and Inanna's journey to her. That myth aids our journey into the dark mysteries of decline and death and rebirth (change), allowing ourselves to be stripped away of our powers and our old identity to face the unknown future.

Hera

Hera is best known as the jealous wife of the Greek King of the Gods, Zeus, always being betrayed, always chasing after him and checking up on him, and often taking it out not on her unfaithful husband, but on the unfortunate object of his desire. There is an older Hera, who may originally have been a Great Goddess with a consort rather than a ruling husband king.

Robert Graves, quoting the ancient writer Pausanias, reports that Hera was worshipped in three forms representing the three archetypes of a woman's life. In the spring she was Hera the maiden (Hera Pais), in summer the adult woman Hera Teleia, and in the winter Hera the widow or crone (Hera Khera). She was especially worshipped at her temple in Stymphalus in the Arcadian mountains.

In her first aspect Hera represented the power of the regeneration of spring. Her second stage was about fertility, motherhood, and the fruits of the field. She also represented a powerful ruler and law giver, wearing a crown. In the third stage Hera went into seclusion, her consort dead, representing the retreat of vegetation and warmth in the winter until the coming of spring again. She was associated with the underworld and with prophesy.

Every year at the end of summer a festival called the Heraeon was held at the time of the full moon. This was similar to the later Olympics. Races were run by women divided into three different age groups, and each group raced with free flowing hair and short tunics. The winners of each group were crowned with olive wreaths and partook of the cow sacrificed in Hera's honour (cows were often associated with great goddesses – see Egypt and India in particular).

You might be aware that within yourself you carry the three aspects of maiden, mother and crone. In what ways do you still feel that the young woman, starting out in life and perhaps free of responsibilities, full of hopes and ambitions, is still within you? Or perhaps it is the mark that life left on her, that remains within you today? In what ways are you still fulfilling the archetype of the mother, strong carer of others but also ruler and lawgiver? Have you yet encountered the crone, beginning to explore the mysteries of the thin veil between life and death, sun and shadow, mundane and sacred? Or are you well advanced on her path?

How might you call on the three archetypes today? There might be the rebirth of new hopes and plans, a resurgence of energy for new projects; the strength and maturity and productivity of the mother, to keep on steadily with what you have resolved; the wisdom and seclusion of the crone Hera, withdrawing from old concerns as you slowly bring about the turn of the seasons and the return of a new, different spring.

Consider the Heraeon, how the women running the races were divided into three age groups. Take this as a metaphor for honouring all three aspects of Hera. We are not to compete with the young maiden for beauty, nor, if retired, to miss the role in the workplace of the mother archetype. Our business is to find out what we are about as a woman at this time of life. We may still be a lover, a mother, a creator and a worker, but the way we carry out these roles may be very different now from how we saw such roles in the past.

We may also take the maiden aspect to encompass the child. When we were children, we were at the mercy of our family and our environment and often didn't understand what was happening to us and why. Our self esteem might have been dented as a result, so that we are unable to fully appreciate our own wisdom, our own beauty and worth as we age. The crone stage may be the time when we really get to know the child within us, as we can use our wisdom and knowledge to understand how we came to be as we are, and what strengths that has given us as well as sorrows. Perhaps now we also use our wisdom to understand and forgive those who had care of us, who did the best they were capable of at that time, but sometimes (or even often) failed us. The child within us can often make us feel like a victim, and an unexamined parent figure within us can be a voice that persecutes us. If we can integrate the child within us we will be healthy and whole.

Hecate

Stories and representations of gods and goddesses change over time, partly to reflect changes in society (as for example the overthrow of female goddesses by male gods), or simply as people expressed different needs in their worship and altered stories to fit them. Hecate is another example of this process.

It is thought that she started out as a goddess of the wilderness and childbirth, originating perhaps in ancient Thrace or Anatolia (on the Black Sea). She was a goddess of liminal places, from the Latin *limen* or threshold. These were places where one area changed into another, such as the shore between the land and the sea, or where cultivated land changed to wilderness. Thus she became a goddess of the wilderness and of changing states. Perhaps this is why she also became a goddess of childbirth, the threshold of independent life for a human being, and the threshold of becoming a mother.

She is known to have been served by eunuchs at her most important sanctuary, which suggests she was an aspect of an old Great Goddess. When she was later adopted by the Greek civilisation in the 6th century BCE she was revered as a mother goddess. Hesiod in his description of Greek gods described her as honoured “above all” by the king of the gods, Zeus, saying she held privilege on earth, in heaven and in the sea. She is described as bringing victory in battle, good luck in gaming, fertility in farming. Although she was now supposed to owe her power to the male God King, perhaps originally she would have had such power by her own right.

There was a common two-way process of empires expanding into foreign lands and importing back into the heart of the empire the gods and goddesses of those new lands. Something like this must have happened with Hecate. One myth told of Hecate as the midwife at Zeus’ birth. Her role as goddess of liminal places led to her image being placed at borders to ward off danger - at the gates of cities, and the doors of households. As well as keeping out evil spirits, she could let them in if offended. Eventually she became seen as Goddess of the border between living and dead – the everyday world and the spirit world.

Through the Greek empire her worship travelled to Alexandria (Egypt), where she was seen as a goddess of sorcery and Queen of the Ghosts. As a result she is often seen today as a goddess of witchcraft. In Egypt she was referred to as the she-dog and associated with the barking of dogs. Later she was shown with two ghostly dogs as servants at her side. Black dogs were once sacrificed to her. The roots of her association with dogs is thought to date back to her pre-Greek days. She was also associated with frogs, symbols of foetus and gestation, due to her role as midwife.

The earliest figurines of Hecate show her as a single figure, but by the late 5th century BCE she was depicted as a triplicate figure, either three bodies together or with three heads. She might be shown as three women or as three animals, such as dog, snake and horse. She was also known as Trioditis (Greek) or Trivia (Latin) – goddess of the three ways. This is thought to arise from her original role as goddess of liminal places, boundaries and borders, and the wilderness. The place where three roads met was representative of the move into unknown and possibly dangerous areas, and sacrifices would be made to Hecate at these points for a safe journey.

Hecate held a flaming torch, perhaps a symbol of the light of wisdom since the Greeks saw her as a bringer of wisdom. Her knife may have represented cutting the umbilical cord; she also carried a rope. The key she carried referred to her role as gatekeeper and opener of doors. She is also sometimes associated with Persephone and thus carried a pomegranate. Many plants such as yew and hallucinogenic herbs were also associated with Hecate.

In the Chaldean Oracles from Alexandria she was associated with a serpentine maze around a spiral, known as Hecate's wheel. Snakes were associated with wisdom, also rebirth (because of the shedding of its skin), and the symbolism of the whole is thought to refer to Hecate leading humankind to life and wisdom.

Hecate was associated with Artemis in classical Greece. Artemis was goddess of the hunt and of the moon, also of fertility (some of her statues are many-breasted). Hecate was worshipped at Artemis' magnificent temple at Ephesus. According to Hesiod (about

700 BCE) Hecate was a cousin of Artemis and their grandmother was Phoebe, another moon goddess. Now Hecate was also moon goddess, appearing in the dark of the moon.

Thus she became a crone moon goddess. Some triple goddesses are presented as Persephone (maiden, waxing moon), Demeter (mother, full moon) and Hecate; or Artemis (waxing moon goddess), Selene (full moon goddess) and Hecate; or Hebe (maiden, cupbearer of the gods, waxing moon), Hera (wife of Zeus, mother, full moon) and Hecate.

Hecate in her cave was the one who heard Persephone's screams as she was abducted by Hades, Lord of the Underworld, with Zeus' approval. When Demeter, mother of Persephone was wild with anxiety over her inability to find Persephone, Hecate suggested that she ask the God of the Sun whom Hecate knew to have witnessed the abduction. Thus Demeter learned the unwelcome truth. Hecate is the goddess of wisdom, of words, and particularly of unwelcome truth.

So Hecate has a number of aspects, many of them dark. She is the midwife of new life but also of perilous changes, and is the bringer of wisdom and unwelcome truth. She is the goddess who presides over boundaries between known and unknown, light and dark, old and new, civilised and wild, live and dead, She has power over the land and the sea, the heavens and also hell. She is goddess of sorcery, of healing plants and hallucinogenic visions. She stands where three roads meet and choices have to be made as to which path to take. She roams in the wilderness and in the dark places with her hellhounds and sacrifices have to be made to win her good fortune. Hecate the crone goddess rules when the moon is waning. Hecate was feared and banned by the patriarchal Christian church but continues to be worshipped by women to this day.

Hecate is the archetypal goddess for ageing and menopausal women. At the menopause we stand on a threshold between the fullness of our middle years, traditionally the mother years – ripe and fertile and productive - and the years of the ageing crone, years that may seem to promise deteriorating health, declining wealth, influence and productivity, with death at the end. Because this time is so new and unknown to us, it is easy to be afraid. We need Hecate's courage and good fortune as we face the unwelcome change to a new state. We need to invoke Hecate to remember that just because the third stage of our adult life is new and unknown to us, that does not mean others have not known it and successfully negotiated it. The threshold we are passing over has been passed by our sisters without number throughout the existence of the race.

Thresholds, liminal spaces, are scary by their nature. They are also magical places, and that is why they have a goddess of their own. Liminal spaces follow different rules and time is experienced differently there. A liminal space may be a time of suffering or impatience or apathy, but it is also a space of enormous potential and regrowth.

The third stage of life is Hecate's, the waning years, the crone years. Hecate reminds us that woman is strong and powerful in all stages of her life – young, middle aged and old. We find different aspects of ourselves as we age, but one is not more or less important than another. It is simply that humans find it hard to remember that change is the basic nature of life itself, and we always seem to prefer to stick to what we know. Hecate can help us to come to terms with our new way of being – to see ourselves as powerful goddesses to be worshipped and celebrated as we embrace our crone years and strive to grow in personal power and wisdom and courage and knowledge.

Hecate encourages us to face unwelcome changes and unwelcome truth. This is a source of great power. Anyone can run after easy truth; unwelcome truth gives far greater rewards when we have the courage to face up to it. We cannot claim the power and wisdom of age unless we accept the less welcome side of ageing. Many people seek to deny their ageing. Some pursue eternal youth through the advertisements for cosmetics and clothes, botox and surgery. Others distract themselves and ignore the advancing years, and carry on as it there were no threshold. Some become depressed, so that while they avoid facing up to their new state they are unable to live fully or feel any enthusiasm. Grasping the truth firmly avoids depression and time-wasting distractions, and like the proverbial nettle, when the truth is firmly grasped it ceases to hurt. Facing unwelcome truth brings us into the fullness of Hecate's wisdom. We may also have to accept that the truth is also unwelcome to others.

Hecate is goddess of the choices we have to make as we stand at the different paths leading off into the distance. How are we going to live now we are older? Are we going to carry on as before? Start acting old before our time? Try out new things we never had time for before? Stop doing so much and try out just being instead? How long can we allow ourselves to stand at the crossroads with Hecate instead of charging off before the way ahead becomes clear?

One myth of Hecate, perhaps from her time as an aspect of the Great Goddess, talks of her appearing as a boar to kill her lover/consort/son; she then waits for the full moon to restore him to life. This reflects the old myths which embodied the cycle of the seasons and the disappearance then reappearance of crops. It also reminds us that the crone moon goddess still contains the other two, represented as lover and mother. What we are becoming is something more, not something less. We can still have great loves, if we want; we can still mother, our own or others' children. We can also still come into partnership with others and bring forth creative projects of our own.

Hecate is a midwife. What will she be midwife to, for you? What will your path lead to, what new woman will be birthed? Whatever happens, you won't stay the same, no-one does. Hecate will help you give birth to your creativity in whatever form it takes, she will make it as painless as possible, or at least help you to triumph as the birth takes place.

Hecate is the goddess of the wilderness, of the wild places. She is associated with hallucinogenic plants, with witches, sorcery, the underworld, spirits. Now is the time to be as wild as you wish – let your imagination run free, cast off the conventions. You are in the third stage of your life and you are increasingly aware that time is limited. Now is the time to dream dreams and see visions, to take the road less travelled. Remember sacrifices may have to be made to follow the chosen path.

Hecate is the goddess of the dark places, the waning moon. If your journey is taking you through your own underworld, perhaps through fear and despair, regrets and bitterness, memories you would rather had remained hidden, truths you don't want to know, Hecate is with you. Perhaps the changes you face seem to have no redeeming features, such as a diagnosis of some incurable illness or disability. You have to travel through the dark place to acceptance and thus to a recognition of new possibilities. While you are in the dark it seems as though it will never be light again, and indeed it is impossible to say how long any dark night of the soul will last. It could be days, weeks, months, years. If we knew how long exactly it would last then it wouldn't be a dark night. Better by far to travel on through the darkness with Hecate at your side than refuse to travel at all (by depression, or denial, or suicide). Hecate reminds us that we are not alone in the dark – many, many of our sisters have travelled the same road, are travelling it now, and will travel it in the future. We are together even though we can't see each other. When we remember that we can bless each other.

Ereshkigal

In the religious myths of ancient Sumer (modern day southern Iraq), the earth was a flat disc surrounded with a rim of mountains floating on an ocean of sweet waters. Above the mountains and atmosphere (lil) was the sky vault containing the sun, moon and stars. Under the earth was the netherworld where the spirits of the dead lived. The relationships between the gods and goddesses changed through time and were often dependent on the prominence of the cities to which they were linked.

Inanna (Ishtar in the Semitic language Akkadian) was goddess of the city Uruk. She was linked to Venus, the morning and evening star and was goddess of carnal love, of war and of fertility. She had many lovers, led armies into battle, and was both beautiful and voluptuous and angry and formidable. Enki (Ea in Akkadian), of the city Eridu, was god of fresh water and wisdom, revered as the inventor of science and the arts. He was deemed to hold the *me*'s, the key powers and laws of nature and civilisation, and to have put the world in order. Inanna was given the me's by the god Enki, and was thus a powerful deity.

In one myth, Inanna decided to visit Enki at his holy shrine in Eridu and he met her with food and beer. As they drank together they toasted each other, and Enki was moved to generosity, giving the me's to Inanna. Inanna set off in the Boat of Heaven to take the me's back to Uruk. When the effect of the beer wore off, Enki realised what he had done and tried to get the me's back, sending monsters to seize the boat. Inanna's counsellor Ninshubur warded them off, and Inanna got safely back to Uruk with the me's.

She loved the god Dumuzi, who was protector of herds and flocks and the god of vegetation that dies in the summer and revives in the spring. There was a ceremony each year, on New Year's Day, in which the king played the role of Dumuzi and consummated a sacred sexual ritual with Inanna, played by her priestess. Only by carrying out this sacred marriage would the return of fertility to the land be guaranteed, and the ritual validated the king's rule. One royal hymn dates from just after 2,000 BC and specifically refers to the sacred marriage of King Iddin-Dagan. Inanna's eager sexuality was celebrated in these hymns.

The Sumerian underworld or "land of no return" was a shadowy place somewhere underground, with a huge palace ruled by Ereshkigal, dark sister of Inanna. After death the spirit of the dead had to cross a river by ferry (as did the Greek dead), and take off their clothes. It was not clear what sort of place it was for the dead spirits to exist in, but from the grave goods now found and the fact that there was a judgement, it probably varied according to how good you had been, and how much your relatives were able to send with you on your journey. No doubt the kings expected to live in luxury there.

Originally Ereshkigal ruled on her own, served by a number of minor deities and spirits, but according to myth the gods sent Nergal, god of war and pestilence, to visit her. When it was time for him to return she threatened to revive all the dead and send them back to the surface of the earth unless Nergal was sent back to her as consort.

One of the great myths of ancient Sumer is the descent of Inanna, Queen of Heaven, to her sister's realm. In one version she goes to extend her own realm, in another version she is invited down for funeral rites. Whatever her motive, she took her seven *me*'s (the sacred powers), dressed and adorned herself, and called her faithful servant Ninshubur to her. She obviously had an uneasy premonition since she told Ninshubur that if she did not return, Ninshubur was to set up a hue and cry. Rather than let Inanna be put to death she should go the great gods, and let them know her danger. Ninshubur promises to see to it.

When Inanna arrived at the outer gates of the underworld she knocked on the door and demanded admittance. The gatekeeper, on the orders of Ereshkigal, demanded that she shed one of her garments before she passed through the door. Presumably her garment represents her powers, one of the *me*'s. Inanna had to pass through seven gates altogether to reach Ereshkigal, and at each gate she had to shed another garment. By the time she arrived she was naked and powerless. Ereshkigal promptly killed Inanna, and hung her dead body on a hook to rot.

After three days and nights when Inanna had still not returned, Ninshubur went to the god Enki and begged for his help. Enki fashioned two demons to go down into the underworld and find Inanna and bring her back, which they duly did. However, the immutable law was that Inanna could not return permanently unless she could find a replacement.

On Inanna's return she found that Dumuzi, her consort, former shepherd and King by her validation, was sitting on the throne, quite content to rule in her place and showing no signs of grief. In anger Inanna handed him over to the demons to be taken to the Netherworld. After a while Inanna missed him and was also moved by the sorrow of his sister, Geshtinanna, and agreed that Geshtinanna could take Dumuzi's place for six months of the year.

This is another myth that explains the turn of the seasons, why there is a loss of fertility and abundance followed by a return each year – because Dumuzi, god of fertility, is missing for those months. There are echoes here of the Greek myth of Persephone being carried off by Hades and eventually having to stay in the underworld for four months a year. But the differences are marked – in this myth, the ruler of the underworld is female (Ereshkigal), and the helpless one is male (Dumuzi), while it is the goddess Inanna who decides who has to be sacrificed, not the king of the gods.

Think of the courage Inanna showed to undertake the journey down to the netherworld. Clearly she knew this was dangerous, since she warned her servant what to do if she didn't return. This journey into the underworld can symbolise our willingness to face up to death, perhaps by boldly facing the fact of our own eventual death, made more starkly clear to us with the arrival of the menopause and other signs of ageing. Perhaps it can symbolise the acceptance of the death of certain hopes, dreams, ambitions, projects, which we must now accept will never happen.

Those who have never had children may need to mourn the end of fertility, whether or not they ever wanted to be a mother. We may have had many plans that have never come to fruition, perhaps because we were not able to make them come about, or perhaps because we put them off for too long and now it is too late. Facing up to this loss involves a kind of grieving. If we allow ourselves to face it, we will come through it and be healthier, more at peace, and more ready to make good out of what is rather than what might have been.

Facing up to all these things is not necessarily easy, and if we get stuck in our grieving then we may need help from our own equivalents of Ninshubur and Enki. Maybe it will be a friend or group of friends or a self-help group, or a counsellor or spiritual director who will help to get us back on the path. Socrates said the unexamined life is not worth living, and it is a fascinating journey to explore our shadow side. But there must always be a return to the external world. We have to balance looking within to turning outwards. We need to be aware if we have stayed down too long.

Inanna allowed herself to be stripped at each gateway, until she had lost everything and was helpless and unable to protect herself. Ereshkigal struck her dead and hung her on a hook as a piece of rotting meat. As we age, we have things stripped away from us – we saw earlier how our very identity may change, sometimes so swiftly that it is painful. We may feel so stripped away of what we thought we were, or of the people around us, or health or circumstances, that we feel as if we are hanging on that hook. Sometimes the only answer is to hang there for a while until life returns.

Inanna and Ereshkigal were sisters, and can be seen as two sides of the same person. They may represent our persona, the person that we like ourselves to be, and our shadow, the parts of ourselves that we keep hidden from view, perhaps even from ourselves. Using the symbolism of the myth of Inanna you can acknowledge the shadow side of yourself, the part of yourself that is Ereshkigal. Only you can say what that is. But by integrating the shadow with the persona, you are a much richer, wiser and more complete and whole person; without the shadow, we are lacking in many strengths and qualities and thus much shallower. The world is not all sweetness and light, and part of the wisdom of ageing is not only knowing that but accepting it without fear or bitterness.

For me the myth expresses how it feels at times to be caught in the underworld; also to celebrate the courage of women who are willing to explore the dark side and the wisdom and richness of character they gain by doing so. It contributes to the sardonic humour so often experienced with older women, who accept things as they are and still say yes to life.

CHAPTER 15 – The Dark Side: Menopause, Sickness and Death

Of the hundreds of useful books on the subject of the menopause that can be bought, or easily borrowed from the public library, one that I would recommend *Is It Me Or Is It Hot In Here* by Jenni Murray. This is a good, readable summary of what happens, with stories about different women's very different experiences, the pros and cons of HRT etc. My only gripe is that because she had no sexual problems at the menopause she seems unwilling to believe they really exist.

My second suggestion is *The Change* by Germaine Greer, which though longer and denser is richly rewarding and worth reading more than once. She looks at the menopause from a much wider perspective than the physical alone, although this is treated in detail and I was reassured to find various odd symptoms in there. However she also looks in detail at the history of women's experience of the menopause itself and their experience of others' response to them, particularly the medical establishment. Much of her material comes from women's own writing. While acknowledging the wide range of experience that different women can have, she sees the climacteric (her preferred term) as a time of emotional and spiritual transition as much as a physical one.

I have also found the internet to be a useful source of information (eg power-surge.com, minniepauz.com) – google various key words and you should find a variety of sites. On ones with message boards you can read directly of other women's experiences, and it is always comforting to find that you are not alone in your particular symptoms. It's also good to read the jokes and dry humour often found on the sites – laughter is healing, but some jokes are not funny coming from anyone other than your sisters in menopause!

It can be quite hard to separate out ageing and the menopause in terms of its effects on the body and mind. We might blame the menopause for symptoms that would have happened anyway. I find that doctors have their own, very short list of symptoms but by reading of women's own experiences you can gradually find that the true list is long. This is not to say that all women experience all symptoms; some women might have only one or two symptoms, another might suffer terribly, most will have their own interesting but not unbearable constellation of symptoms.

It seems to be the case for many women, and certainly was for me, that the symptoms feel more distressing when you don't know what they are, or think you are the only one suffering them, or fear that they are symptoms of some terrible underlying illness or permanent disability. Once the symptom is recognised as a consequence of the menopause, shared by thousands or millions of other women, and likely to subside over the next few months or year or two, then it is endured with reasonable resignation. This is one reason why it is so important for women to share their experiences with each other.

However, the older we get, the more likely it is that illness or accidents will occur and will alter who we are. The picture we have of ourselves includes our physical abilities – if these change, we have to adjust our identity accordingly and this can be painful at first as well as frightening. It is particularly difficult when we have an abrupt change. We may be able to cope well with on ongoing, gentle ageing process, but a sudden discontinuity in our physical abilities can be very hard to adjust to. How could we suddenly change from one person (fit and unimpaired) into another (damaged)?

There may then be major changes in life circumstances, such as loss of income leading to poverty, or increased dependence on other people as physical capabilities are lost. Or the changes may be invisible to other people, but still requiring painful adjustment to a new situation. Change will be particularly difficult in a society where anything different from an idealised norm is not acceptable. It helps to share your story with other people – you will soon find that the "norm" is not that normal!

Kathy Charmaz interviewed a number of chronically ill people. She found they often restricted themselves more than was necessary, due to fear and uncertainty about their condition. Feelings of not being in control, of having little information about their condition, made things worse. Constantly thinking about what was wrong narrowed their horizons. This in turn increased their social isolation.

In some cases the effects of a condition caused other people to relate differently to the ill person. Slurred speech or deafness might mean they were treated like a child or an idiot; invisible symptoms may be seen as hypochondria. Being patronised or otherwise treated differently might be shocking to the ill person and attack their self-identity. And some lost their self esteem as they felt they no longer had much to offer.

Even uncomplicated ageing can cause such difficulties. A woman going through the menopause may find that as she visibly ages, people in the street treat her differently, and she may feel less visible, less respected. She feels that other people see her as a "silly old woman". This may be a shock if she feels pretty much the same inside. Finally if she is feeling wretched, with less energy, and perhaps not fulfilling various roles that other people have come to expect from her, she may come to feel that she is a burden on her family and community.

These are situations where the ability to draw on feelings of the strength and courage of a Hero rather than a Victim are very important, whether we are talking about a major illness or disability or comparatively minor changes of the menopause. Of course we must allow ourselves some time to feel hurt, shocked, bewildered, or sorry for ourselves when unwanted changes occur, but we cannot afford to remain in that state all the time.

Some feeling of control can be regained by gathering all the information we can. The internet is a very useful tool, but we may also need to be firm with health professionals. Taking someone along as an advocate at consultations can be helpful here. If other people treat us with less than respect, we need to point this out – it's a good education for them as they may be totally unaware they are doing it.

It helps to mix with people who are in a similar state to us, partly to feel more "normal", partly to gain information and be inspired by what others can achieve. Self-help groups are important here. If there isn't one in your area, why not start one? Be prepared to accept help from others without letting it make you feel bad. Even heroes accept help.

It is important for a woman to be clear about what is reasonable for her and what is not. She may well feel a need to withdraw for a while; this needs to be framed as a positive move toward caring for herself and her journey rather than a negative. It would be sensible to let those around her know in brief what is going on, so that friends do not feel simply ignored and neglected and will be ready to take up the relationship again when the woman is ready. It may also be wise to keep some contact, perhaps occasional telephone calls; and to have one or two companions on the way.

To maintain a comfortable self image, and to be able to value her new way of being rather than see herself as useless or a burden, the woman needs to find positive images to relate to and to shield her from society's negative stereotypes. The more that women in and beyond the menopause claim the right to honour their physical changes and to exchange information and ideas, the less marginal they will feel. However, the sense of self identity will still change, and will need adjustment.

Remember life is a journey, and no one journey can be rated as more important than another. I remember reading somewhere that "God is more likely to work in you than through you." We'd all like to make a difference, but sometimes it's hard enough just to be ourselves. And be kind to yourself – if there are days when you collapse into victimhood, you can always start again when you feel better. Learn when you need to be firm with yourself and when you just need to let the emotions wash through you. Even heroes have to lie down and weep sometimes.

A journey through the changes in self identity due to serious illness is described by Joan LeMaistre in her book *After the Diagnosis*. Dr LeMaistre was diagnosed with multiple sclerosis just after receiving her Ph.D. in clinical psychology and giving birth to a daughter. She worked with chronically ill patients and their families for many years. With similarities to the idea of the stages of grief, she outlines a series of stages that a person may undergo as they adjust to the changes in their life: crisis, isolation, anger, reconstruction, intermittent depression, and renewal.

First there is the initial crisis and accompanying emotions of fear, devastation and loss, anger and a sense of injustice. This can take a long time to get through. Later, reconstruction can begin as the patient starts to set herself goals and learn new skills, with perhaps new social contacts and new interests. It is not a return to the same life as before; there is a new life and a new self. LeMaistre sees this as the reemergence of a new, positive self-image.

However, the person needs to be aware of the possibility of intermittent depression and to be ready for it. The losses do not go away and sometimes may be felt as if new. LeMaistre talks of the 'phantom psyche', the image of how you would be without the illness; the you that you used to be. This is usually felt in terms of "if only" – "if only I didn't have this arthritis, I could still be doing what I used to do". The phantom psyche may gradually diminish, but if we accept that there are times when we have to grieve, it will pass more easily.

Finally there is renewal. Although the losses, and accompanying sadness, may never entirely go away, the person accepts the illness even though they don't like it. There is personal growth, a new future planned, new options created, new expectations and hopes based on things already achieved.

Whereas LeMaistre's book is aimed at people with a diagnosis of a severe illness, the stages she outlines may be useful for all sorts of changes in self identity. Luckily for our continuing sense of a stable self, change is usually slow enough and minor enough for us to adjust without particularly noticing it. Major changes, even happy, looked for changes such as childbirth and marriage, moving to a more desirable house, starting a better job, can all make us feel undermined for a while. The changes we undergo through ageing can be stark enough to shake us up considerably. If we bear that in mind, we can be kind to ourselves and give ourselves plenty of time to adjust.

The conscious ageing movement believes that modern Western society fears and denies ageing because it fears and denies death. Rabbi Schachter-Shalomi believes we need to come to terms with the personal inevitability of death. He believes that we cannot carry out the tasks of ageing if we expend too much energy on denying our mortality. Ram Dass believes we should be as conscious as possible of what is happening to us, whether during ageing or at the point of death, as an essential part of our spiritual journey.

Humans are the only species that knows it is going to die. We cope with that knowledge mainly by denying it. Most of us are mentally defended against thinking too deeply about death. When the fear does emerge, it is usually displaced onto something else. When death cannot be denied, there are elaborate religious rituals to cope with the anxieties that accompany it. In our society, death is not so much a part of life as used to be, being tidied away into hospitals and hospices. Very few of us die at home anymore, and we tend to live longer.

Epicurus, the Greek philosopher, said that the fear of death was illogical. We are not bothered by the nothingness that there was before we were conceived and born, so why worry about the nothingness after death? Lucretius also dismissed death by saying that where death is, we are not; where we are, death is not. In other words we don't need to worry about death because we won't be there to experience it.

It is perhaps nothingness and not being here that we most fear. However, it is true that Epicurus' take on death is supremely logical.

Thus death anxiety must be rooted in our feelings rather than the analytical part of us. The intellectual knowledge of our approaching demise is due to the reasoning power of our brain's frontal lobes. The fear that we feel is due to the workings of a much more ancient part of our brain, the amygdala, and accompanies any threat to our existence. The two become connected in death anxiety.

Part of the impulse to feel fear from the existential knowledge of our own mortality probably comes from the ego, the part of us that holds our working memory – those memories that we have conscious access to at any one time. When something reminds us of death, we may consciously feel fear, or repress the thought to avoid discomfort.

The ego's job includes maintaining our sense of self, which it equates with itself. The ego cannot stand the thought of its own annihilation, becoming nothing. The ego equates its own absence with the death of the whole organism, and resists it fiercely. That is one reason why it can be so difficult to meditate. You may feel that if you actually managed to stop the ever-flowing stream of thought after thought, you would cease to be able to function. When experienced meditators or yoga practitioners do manage to transcend the ego and the endless chatter of its thoughts, they report that death ceases to be something to be feared.

In his book *Beyond Dying*, journalist Ted Hanson talks about the Near Death Experience (NDE), an altered state of consciousness where something seems to be experienced beyond death, though the patient is then "sent back". Scientists such as Michael Persinger and Susan Blackmore see NDEs as a result of a breakdown of the internal model of self, due to the collapse of normal brain processes. There is the feeling that a separate "I" never existed, and therefore "I" cannot die. Hanson himself is unwilling to let go of the possibility that the experience is real, in spite of medical explanations. He also looks to other out-of-body experiences, reincarnation and "psi" phenomena as possible manifestations of a soul that is separate from the body and that can thus survive death.

Jiddu Krishnamurti says that it is the word, the image of death that creates fear. We cling to the known and fear the death of our limited self and the loss of all the "things" that go to make up our image of our self, which is actually composed of memories – the dead past. This is our ego. If we can learn to die psychologically to each moment, instead of clinging to the security of the past, then not only is life fresh and beautiful but our fear of death will cease.

Mystics from all religions may spend time meditating on death, from the desert monks of Christianity to the Buddhists who sit meditating next to corpses. Mastering the deepest fear of the future can help us to be fully in the present, to move beyond all other petty concerns, to get a sense of what is truly of value, to feel and love and even to think more deeply. People who have had brushes with death often talk about their realisation that they were not spending enough time on what really matters, such as loved ones, and that their values had altered.

I have found that there is a new poignancy to spring now that I am post-menopausal. I think this is largely because I am aware that there is a limit to the number of times I will see such a wondrous process take place before my eyes; there are fewer springs ahead of me now than there are behind, even if I live to a ripe old age. As Germaine Greer says:

> "The feeling that one's day has passed its noon and the shadows are lengthening, that summer is long gone and the days are growing ever shorter and bleaker, is a just one and should be respected. At the turning point the descent into night is felt as rapid; only when the stress of the climacteric is over can the ageing woman realise that autumn can be long, golden, milder and warmer than summer, and is the most productive season of the year. The elegiac strain never fades quite out of the middle-aged woman's consciousness, but it gives poignancy to the now rather than the bitterness of regret that is felt by some so keenly at menopause. When the fifty-year-old woman says to herself, 'Now is the best time of all,' she means it all the more because she knows it is not for ever." (Greer 1991 p142-3)

The German existential philosopher Martin Heidegger noted that the thought of death produces a feeling of dread ("angst") in us, when we truly feel that "I must die" rather than "one must die". That existential anxiety can be masked in a state of busy-ness and by an illusion of control over our daily lives; an actual brush with death can upset that illusion. Heidegger felt that to live authentically we must construct our lives in the full knowledge of our eventual death, otherwise we will not exist fully and truly. Full consciousness means full self-awareness, including our awareness of death. Being half asleep is no answer.

Being made more conscious of our own death by the onset of menopause may result in a fully fledged death anxiety, or a mourning for all that is lost to us – youth, looks, power and status, general potential and choices, way of life, perhaps actual loved ones. We may feel a gentle melancholy as we watch the seasons pass, while philosophising on the meaning of it all.

However it manifests, it is far more difficult for a woman to avoid consciousness of our limited lifespan than for a man because the menopause is so sudden and in a sense, so ruthless. We may not welcome such a process, but we must surely gain something from it. Saints and mystics, philosophers and others searching for their authentic self work hard to reach insights that should come much more easily to us. We are pushed into their hard-earned insights by Mother Nature; she turns us into the Wise Woman. However we still need courage to face up to these things.

Try constructing a spider diagram. Write the word 'death' with a circle around it in the middle of a piece of paper. Now draw spokes coming off from the circle and draw more circles on the end of them; inside the circles, write anything that pops into your mind in association with the word death.

What are your particular thoughts, fears, hopes around death? What is the history of your attitude to and thoughts about death? What did your family teach you about death and how to react to it, even though it may have remained unspoken? What have you absorbed from your culture? Which beliefs do you encourage in yourself? How do you deal with thoughts of death? Have you had times when you have felt in touch with something beyond this existence? Is there a difference between what you think is true and what you feel is true?

Conscious ageing is a life choice made in the belief that it will improve the quality of life. If there is too much death anxiety this will instead be counterproductive and lower the quality of life. Therefore the next chapter examines ways in which to reduce death anxiety.

CHAPTER 16: The Dark Side – Coping with Death Anxiety

In his poem *Aubade* Philip Larkin talks of the death anxiety he experiences in the dark of night. As he says, most things we worry about will never happen but this definitely will. Death anxiety can be a very unpleasant experience, and doesn't necessarily only occur in older people. We need to be able to accept the idea of our own death without letting this ruin the life we are still living.

For people whose religion or belief system includes a belief in an afterlife, the idea of death may still be alarming but they will not have that feeling of being faced with annihilation that other people may have. I describe myself as a Christian, but do not believe in heaven or hell or miracles (or even a bodily resurrection). How might we non-believers cope with the idea of death?

The astronomer Carl Sagan was a man who always showed a dedication to the truth. In his last book, he wrote about the illness that killed him – the initial realisation that something was wrong, the diagnosis, the course of his illness, the hopes of recovering after a transplant, the disappointment of a further downward spiral. He died before the book was published, and his partner added a chapter on his final days. Carl Sagan remained an atheist and faced his end with great courage, taking comfort in the love of his friends and family. He left children, books and the legacy of his work behind him, and must have had a sense of a life well lived, especially since he was appreciated and honoured in his own lifetime.

Irvin Yalom is an existential psychotherapist, and as such is interested in the great givens of human existence – one of the main ones being the fact that we have to live with the idea that we will die. He sees one of our major developmental tasks as the need to cope with this knowledge, and believes that the fear of death is ever present throughout life even if hidden from our consciousness. Nightmares are one way in which it escapes. He ran group therapy for terminally ill patients for many years.

Yalom explores death anxiety in his book *Staring Into the Sun.* He believes that facing our own mortality brings us insights that enable us to live more fully and authentically. "Though the physicality of death destroys us, the idea of death may save us." The jolt that the fact of death gives us – perhaps knowledge of our own death, or experiencing the death of someone close to us – takes us out of the mundane everyday state of existence and gives us a different perspective on life that may be life-changing and lead to a wiser, more authentic, more fully lived existence.

Yalom sees two major ways in which we deny death: one is the belief in a supreme "ultimate rescuer", such as God, the other is our own "specialness" – the idea that whereas everyone else must die, somehow we won't. He himself occasionally suffers from death anxiety at night but comforts himself with Lucretius' doctrine (where death is, I am not). An atheist, he sees his death anxiety as a price worth paying for the deeper sense of authenticity and meaning that it gives to his daily life.

Sheldon Kopp (author of *If you Meet the Buddha on the Road, Kill Him*) was another existential psychotherapist, deeply interested in spiritual matters and a believer in God but not religious in a conventional sense. He underwent brain surgery in middle age and nearly died, though he survived incapacitated until about 70 years old, writing many more books and articles. Kopp chose to see himself as having been handed the task of finding meaning in the dreadful experience of his illness. For Kopp, facing up to our fears is the only way forward to a free, authentic life; in the face of death, the answer is to live well.

Ageing consciously means living fully and authentically without repressing the inescapable knowledge of our death. We should gain a more meaningful perspective on our lives, with a better sense of what matters and what doesn't. The days left to us should have a sweeter poignancy for knowing that they are not inexhaustible. However, we should do all we can to avoid tainting our days and nights with avoidable death anxiety. Some comfort may be found from the following ideas.

Religion/Spirituality

Throughout the ages, organised religion, superstition or a more diffuse kind of spirituality have been used to cope with the knowledge of our mortality. Even for non-believers, there is no reason why some sort of spiritual approach should not be used. Indeed, many people are spiritual without even realising it.

We all have the task of constructing meaning for our lives, though most of the time we do it so easily that we are not aware of doing so. The idea of the holy, the numinous, is found throughout human cultures, and seems to be an innate aspect of the way our brains work. Through spiritual practice one can learn to diminish the grip of the ego and the sense of being separate, and increase the feeling of unity consciousness. The less we are concentrated on the idea of a separate "I", the less there is to fear from death. We are only part of the whole, and as part of the whole, all cannot be lost.

For Jung the purpose of religion is to give expression to the religious archetypes and symbols preexisting in our brains. Since these are beyond words, and are a facility or predisposition for religious feelings and ideas rather than a religion in themselves, no religion with its dogma and rituals can be the one true religion, but only an attempt at the expression of an ideal. However as Sogyal Rinpoche says, it is best to devote ourselves to one spiritual path, rather than to take a pick and mix approach, as otherwise spiritual progress will be difficult. It is best to choose one wisdom tradition and grow in that.

Aldous Huxley talked of the "perennial philosophy". This is the set of ideas that is common to all religions in their mystical form. Practitioners aim to reach an altered state of consciousness, often by setting aside the normal chatter of the thinking brain through prayer or mediation, and get in touch with some eternal sense of meaning, oneness and sacredness. Whatever the religion, the experience, once beyond words, seems to be similar – an outpouring of the human brain, not culture. Compassion, integrity, truth, love, wisdom, all seem important; material considerations such as fame and money seem trivial or counterproductive.

Timothy Ferris in *The Mind's Sky* suggests that what happens when we have such a peak experience, is that we break through the common workings of our mind to an "integration programme" which lies underneath, so to speak . He theorises that there is so much going on the brain, conscious and unconscious, different "modules" as he sees it (including a language module), that this integrator is necessary to give us the illusion that the mind is a unified whole. Mostly we are unaware of it – until we access it directly and have the feeling that literally all is one. And we feel the experience to be beyond words because we have, literally, gone beyond words.

Whatever the explanation, and whether you are religious or not, it is possible to have such an experience with or without prayer or meditation. It may be when seeing something beautiful, or hearing some music, or looking at a loved child, or recovering from illness or danger, or simply when walking down a busy street, as happened with Thomas Merton. Suddenly everything, while staying the same, seems to have a completely different significance; all seems beautiful and meaningful and beyond words. There may be a sense of unity consciousness. Religious people are likely to experience religious imagery. The philosopher Blaise Pascal had such an experience in 1654, and when he died a brief outline of it was found written on a piece of paper kept in his waistcoat:

> "From about half past ten in the evening until about half past twelve. FIRE. God of Abraham, God of Isaac, God of Jacob and not of the philosophers and savants. Certainty. Certainty. Emotion. Joy. Peace." (Winston 2005 p 272)

There is also the sense of the numinous, a word suggested by Rudolph Otto to convey the feeling of awe and wonder that accompanies the "idea of the holy" (the title of his book), that might overcome any of us given the right setting. Cathedrals were built in order to inspire this feeling – even non-believers may feel constrained to talk in hushed voices as they wander under the great arched pillars lit by the dim coloured light of vast stained glass windows. It gives us the sense that certain things demand our respect, even if they are symbols of a religion alien to us. Nature can arouse this feeling.

The sense of meaning and the sense of the holy often go beyond words although they all require the modern human brain. There tends to be an emphasis on language among brain scientists which can overlook other ways in which symbols are used, and there has been an over-use of the analogy of the brain as a computer, with the emphasis on logic. Certainly the use of logic and language which has been made possible by the evolution of our relatively massive brains is of enormous significance. But they are not the whole picture.

Lloyd Geering in *Tomorrow's God*, a postmodern look at the Christian religion, says that meaning is about language and therefore about intellect and logic. He also says that the first product of a child is language. It seems that once again the stereotypically "masculine" qualities of the brain are being emphasised at the expense of the "feminine" skills of relationship. The first product of a child is not language, but relationship. Long before a child learns to speak, it is in relationship with its mother. Our most basic psychological attitudes are laid down long before language, which is why many psychotherapists place more emphasis on the therapeutic relationship than the topics spoken about within it. Our basic feeling of whether we can trust the world or not is learned at our mother's breast, as is our basic sense of ourselves as integrated personalities.

Our social/emotional side is as much a part of the wonderful human brain as is the ability to speak and to think logically. We cannot ignore one, or the other. We need to honour our need for symbol, meaning and some fulfilment of our religious archetypes as much as we need to use our ability to think sensibly and with an eye to the scientific evidence. Again it is a question of integration.

The psychotherapist Scott Peck in *Further Along the Road Less Travelled* explained his four stages of religious development (there are other schemes). First is Stage One, the "chaotic/antisocial". These are people he found to be lacking in self-discipline and principles, too needy and self-serving to be capable of proper relationships with others, sometimes to be found in jail or mental hospital or on the streets. These people may convert to Stage Two, the formal/institutional, or people may be born into it and not progress. People in this stage depend on an institution for a sense of internal organisation, and religion in Stage Two is likely to be very rigid and formalistic. Their dogma will be set out as a set of iron rules, and their God is likely to be a "benevolent cop in the sky", ready to punish wrongdoers or reward the good people who stick to the rules.

Scott Peck suggested that children raised in a loving, stable Stage Two home will grow up with a strong healthy ego and sense of self, with a good intellect coupled with an easy self-discipline. They may well grow up to think that Stage Two religion is all a load of superstitious nonsense, and want nothing to do with it. They move into Stage Three, the sceptic/individual stage. Sometimes however they might feel that there is something more to life than what they can see, and they might move on to Stage Four – the mystical/communal. They learn to see things in terms of interconnectedness and also of paradox; that there are different types of truth, and that opposites might both be true at the same time. This is the stage of the perennial philosophy, the heart of all religions, where absolute truths are beyond words and have nothing to do with dogma.

When we get beyond words, there is silence. When Yahweh came to Elijah on the mountain of Horeb, he was not in the mighty wind, nor the earthquake, nor the fire, but in a still small voice; or, as in the New International Version of the Bible, in a gentle whisper (1

Kings 18:12). I read somewhere that a literal translation of the original Hebrew actually means "in a silence so profound that it spoke". In Psalm 46 we read "Be still, and know that I am God".

By stilling the chatter of the ego, by transcending the ego, going beyond its necessary but mundane concerns, we can get in touch with something beyond the material truths, whether or not we have any time for organised religion. This is not to ignore the importance of words, but to recognise that there are times when we need to go beyond them. As James W Jones said in his book *In the Middle of This Road We Call Our Life*:

> "Beyond the categories of emptiness and fullness, the sacred is not just pure vacuum. The Divine expresses herself through a variety of images without being identical with them. The speculations of the Upanishads and the image of the drop of water vanishing into the sea, the koans of Zen Buddhism and the lure of paradox, the heroics of Camus and Sartre, the soaring calculations of Einstein whose God does not play dice, the grace of Jesus who forgives our trespasses, the love of the Great Mother whose wisdom is read on the wind – all are permutations of the presence of the void." (Jones 1995 p181)

There are many ways in which to honour our sense of the numinous. As well as timeless symbols and archetypes found in the world's great religions, we may have our own images, art, music or other experiences. Nature is traditionally an important source of religious experience: look at the poetry of Gerard Manley Hopkins, Wordsworth, Shelley and many others.

When we directly encounter nature, we are in what is known as a "thin place" – a place where the Divine underlying our everyday life can manage to shine through. Marcus Borg discusses this idea in his book *The Heart of Christianity*, where he talks of the popularity of Celtic Christianity because of their sense of thin places and their recognition of them throughout their daily round. A thin place may not literally be a place. Jesus would have been a thin place for his disciples, and the Buddha and other holy figures for theirs. You may find your thin place in someone you love, or someone you see as a spiritual person, or in some activity, or symbol. We all need to cultivate our thin places and visit them more often.

A sense of the holy can coexist with intellect and logic and common sense. As well as making life more magical day by day, it can help to defend against the deepest terrors of our death, since it is our ego which most fears death (being the part of us fretting about our future), and the numinous takes us beyond the ego.

What would you die for?

If you would be prepared to give up your life for something, then death cannot be the worst thing that could happen to you. I would guess that the things you would consider worth dying for, in some sense transcend death. Probably most of us hope we would have the courage to die for love: to give up our lives so that our children and grandchildren may survive, or to risk our lives to save our friends and family, or even to risk our lives to save a complete stranger, since we are social beings with altruistic genes in our makeup.

Thus we would risk death for family and community. In that sense we see ourselves as part of something bigger than us, that outlasts us. When we are gone, our family (if we are lucky enough to have one), but certainly our community, and the human race in general, will continue. We have passed on the lamp of life to others: we have had our turn, but others come after us. As part of something bigger than us – the human race, or life itself – we are part of a river of life that continues to flow after we are gone.

As well as being ready to die for love, or community, we might be prepared to die for more abstract concepts: for truth, integrity, justice. People all over the world risk death and torture in order to speak out against oppression and injustice, exploitation and hatred. We recognise them as martyrs to be honoured for their choices, whether or not they follow an organised religion. Again we might hope we would have the courage to do the same, since we can never know unless we are tested.

Some people might risk death for other, less universal reasons: for exploration and adventure, either for a scientific or other productive purpose, or for sport. Such people might say that the freedom and adventure of mountain-climbing or arctic exploration are worth risking death for. Again these are concepts that outlive any one individual's life.

Perhaps there are things that you would risk your life for that are particular to you. Whatever they might be, they all show that there are more important things than death and that they transcend death.

What do you leave behind you?

What will give you a kind of immortality? It may simply be the fact that you have descendants and your genes live on through them. Even if you have no children, relatives will carry your genes. Genetically speaking, you are as closely related to your niece as you are to your grandchild. Perhaps you have produced works of art, or a book, or a business. Perhaps there is a voluntary organisation or political tradition that would not have survived without you, perhaps your contribution is less noticeable but present in whatever you have put your energies into. Perhaps it is about the relationships you have fostered, or the happiness you have brought to someone.

Whatever you have done, your deeds will have their own kind of immortality in the effects they leave behind. Irvin Yalom refers to this as "rippling". The universe is not quite the same place as it would have been, had you never existed. We are all part of the evolution of life, the unfolding of the universe, though it is not given to us to know where it is all going.

Living life to the full

Psychotherapists report that it is people who feel they have only half lived their lives who most dread death, with the feeling that they have been given an opportunity that they have wasted. This is not necessarily about "doing" and "achieving", though that may be part of it. Mostly it is about being yourself, honouring your needs and desires, being the person and undertaking the activities that express who you really are.

When we get older we have more time to do what we want to do, or meant to do but somehow never got round to it. If you always fancied yourself as an author or a poet, an artist or a gardener, now is the time to get on with it. If you have estrangements that you regret, or feel you haven't put enough into certain relationships, it's never too late to try again.

Living life to the full is about living authentically – living the life that you are meant to lead, and not just shaping yourself to please others. It is also about trying to remember to live life in the present, not constantly dwelling on the past or worrying about the future. In one sense the more we can live in the present, the more time we spend alive.

Being more aware of our mortality and the fleetingness of life might bring us all to honour life more than we do, in other people as well as ourselves. We may be more aware of those who are not able to lead fulfilling lives, or whose lives are cut short because of social injustice around the world. Although we may not have the certainties and energy of extreme youth, we may feel able to do our bit towards a fairer world.

Eternal Images

We may comfort ourselves with images, or quotations, or other symbols that for some reason work for us. For me it is a combination of myth and physics. I see an image of life in the universe as a tapestry, such as might be woven by the Wyrd sisters (Anglo Saxon goddesses), in which one of the short threads is my life. Although the thread has a beginning and an end, it is essential to the tapestry and will always be there. This is partly based on my understanding that modern physics says time is simply another dimension. Thus past time never disappears, it is merely unavailable to us, as our third dimension in space would be unavailable to a two-dimensional being. For Irvin Yalom, the comforting image was in a sense the opposite to my image of an eternal tapestry - the words of Lucretius, that there is nothing to fear from death since he won't be there at all.

For others it may be the images of nature – the cycling of the moon, from waxing to full to waning to the dark of the old moon before the return of the new; the cycling of the seasons; the sense of rightness that birth and growth are followed by waning and decay in the great cycles of life. We can think of the old myths of the death of the king in the autumn, the shedding of his blood on the field, with a new king rising again through the Goddess in the spring. We may see ourselves as returning to the womb of the Goddess, as in the old myths. We go back to where we came from.

The image we have of death or dying can make a difference. Seeing death as a step into the unknown can be frightening. If instead we see it as a return to where we all come from, it can seem less threatening. In physical terms it might be the difference between seeing yourself travelling along in a straight line and then falling over an edge into nothingness, and seeing instead a circular journey from the sky or the earth or deep space, into this body and then back to the point of origin. It might be worth spending some time to construct the most comforting image for yourself.

The terms we use will make a difference. Instead of seeing death as "a monstrous thing waiting to pounce on us at the end of life's journey" as Stuart Holroyd puts it, we can encourage kinder images. "Safely gathered in", "returning to the Source", "dissolving into the cosmic ocean", "going to our final rest/bed/sleep" and "going home" are a few suggestions.

We are all different and therefore we will all have our own comfort blankets if assailed by death anxiety. And if you do get an attack in the middle of the night, nothing seems to help and you can't sleep, then look at the ideas in the later chapter for making yourself feel better. If you feel it is unproductive suffering, then just put an end to it – the easiest thing would be to get up, make yourself a hot drink and read or watch TV until you are tired enough to go back to sleep. A general awareness of death may be an important part of conscious ageing, but there is no sense in making ourselves miserable!

If we can see our journey through the last third of our life as the hero's journey into the unknown, on an adventure that may take us into danger and bring us face to face with dragons, it should also grant us boons that we can bring back to serve our community. This journey is into our own psyches, facing up to our fears and repressions (our dragons), and the boons we may bring back include a more integrated self, more self-awareness, a greater ability to live in the now, greater wisdom and ultimately a greater fearlessness.

You might want to try a spider diagram to see what examples of possible boons come to your mind, or perhaps you have already experienced one or more. Some suggestions: self knowledge and growth; mental health; a mature spirituality; wisdom – a stripping away of illusions; independence; to live life to the full; to get on with it now; to be all we can be; if we can accept death we can accept pretty well anything; freedom from fear; the ultimate wisdom of Not Knowing – living with that; a sense of eternity (paradoxically!); new and better priorities in life; new beliefs and perspectives, new goals and activities; less selfishness; a sense of belonging to something greater than ourselves.

Try a visualization of a journey into the underworld. After getting into a state of relaxation, imagine yourself journeying out into the country. You find an arrangement of great stones in a field; within the stones is a trapdoor which you open. Leading down into the earth is a set of stone steps. You descend the steps and at the bottom meet with your Guide. Your guide leads you along a passageway which then opens out into the Elysian Fields, an underworld place of great beauty. Coming towards you across the field is the Great Goddess who encompasses both life and death.

Spend some time speaking with the Goddess (you may need your Guide's help). She may give you words of advice/comfort/wisdom; you may be given a symbol to take back. You might ask – what will help me on my path at this time? What tasks remain for me to carry out while still on earth? You may want to ask the meaning of certain things that have happened or are happening to you.

Now ask the Goddess what she has to say to you about death. What wisdom does she want to share with you? Ask any questions of her, or your guide, that feel you want to.

When you have completed your conversations, leave the Goddess with your Guide and return to the passageway. Ascend the steps and return to the earthly field above. Now retrace your steps back to the beginning of the journey. Start to remember the room you are in, and get ready to return. Start feeling your limbs and fingers and toes – wriggle them – when you are ready open your eyes.

You may wish to draw what you have seen, or make notes; you may or may not wish to discuss it with a partner. The more you express what you have experienced, the more you will draw out of it.

CHAPTER 17: Goddesses for the Dark Side

The old goddesses were often cruel. They did not really fit with the vision of a peaceful, loving, matriarchal paradise. Many were warrior goddesses. Since humans had been created by the gods as their servants, they could expect little from them. The best they could do was to placate them with many rituals, sacrifices, and confessions of sins and failings.

In the kingdom of Sumer and later Akkad, Inanna was Queen of Heaven – but she was about death as well as life. Death was an ever-present reality for pre-technological peoples, and the goddess was seen as all-powerful and terrifying, giver of illness, death and destruction as well as life. Over 4,000 years ago the Akkadian priestess Enheduanna wrote *The Exaltation of Inanna*, in some of the earliest poetry:

> "Queen of all given powers
> unveiled clear light
> unfailing woman wearing brilliance
> cherished in heaven and earth
> …… my queen of fundamental forces
> guardian of essential cosmic sources
> you lift up the elements
> bind them to your hands
> gather in powers
> press them to your breast
> vicious dragon you spew
> venom poisons the land
> like the storm god you howl
> grain wilts on the ground
> swollen flood rushing down the mountain
> you are Inanna
> supreme in heaven and earth."
> (*The Exaltation of Inanna,* Meador 2000)

Early powerful goddesses encompassed all the realities of life – war, famine, disease and disaster, the death of loved ones including children – and were not thought to have a responsibility to make humans happy. Thus instead of railing at Inanna for sending the floods that destroyed the crops and resulted in the starvation of her people, Enheduanna worships her for her power. Death was simply a fact of life. A goddess who is still worshipped in such a way today is Kali.

Through her we can celebrate life in all its energy and creativity while facing up to the inevitability of death and destruction.

Kali

From an early Indian civilisation around the Indus are found terracotta figurines of females and bulls and also phallic emblems, reminiscent of findings in other places such as Catalhoyuk. Merchants' seals and amulets generally depict animals: bull, buffalo, goat, tiger, elephant, and possibly scenes from legends. There are suggestions of a mother goddess with sacred trees and a horned fertility god.

The original inhabitants were conquered by the Indo-Europeans by about 1600 BCE. They brought in their own patriarchal gods and the importance of female deities declined markedly. Indra was a dragon slayer and rider of the storm, and Brahma the creator was another important god.

The Indo-Europeans never managed to completely subdue the south of India, and as a result many of the original deities survived in one form or another, influencing in turn the pantheon of the northern gods and their sacred writings (for example the *Rig Veda* of 1500-1000 BCE, the *Upanishads* of the 1st millennium BCE, and many others).

As time went by, relationships between deities altered and some waxed or waned in importance. Some deities had many aspects and each village had its local god or goddess. The variety of forms of the divine over space and time makes the picture quite confusing. Not only were there anthropomorphic gods and goddesses, and certain animals held sacred such as the cow and the snake, but there were tree cults; sacred hills and mountains, particularly the Himalayas; sacred rivers, particularly the Ganges; sacred lakes and even sacred cities.

Eventually there evolved three main male gods (as well as a host of minor deities and their wives): Brahma the creator (his wife was Sarasvati, patron of art, music and language and writing), Visnu (pronounced Vishnoo) the preserver and Siva (Sheeva) the destroyer. However, many worship only Visnu or Siva. Visnu is seen as the source of the Universe (through Brahma), a benevolent god who sometimes incarnates: as Krishna in the *Bhagavad Gita* and Rama in the *Ramayana*. Visnu's wife was Laksmi, the goddess of good luck and wealth; who was originally said to have appeared in the churning of the primaeval ocean (could this be an echo of an original goddess such as the Sumerian Nammu or Babylonian Tiamat?). Laksmi has many worshippers of her own, and is usually portrayed as a mature and beautiful woman sitting on a lotus, attended by two elephants who sprinkle her with water.

Siva, the destroyer, has elements of the original non-Aryan horned fertility god. He wears a garland of skulls and is surrounded by ghosts and demons. He is death and time which destroys all things. He is Lord of the Dance and his wife was Parvati, daughter of the Himalayas.

The female divine principle was Shakti who is, like Brahma, less anthropomorphised. Whereas the male creative principle was seen as inactive and transcendent, the female element was active and immanent. The male principle was brought to life by Shakti. In more anthropomorphic forms, the great goddess – usually in the form of a mother goddess – is also worshipped. She may be known as Parvati (Siva's wife), Mahadevi (the Great Goddess), Sati (the Virtuous), Gauri (the White One) or Mata (the Mother). In a darker aspect, she is Durga (the Inaccessible), Candi (the Fierce) or Kali (the Black One).

Kali is often depicted as the consort of Siva, equally dark and wild, and they may be shown dancing together or sometimes she is dancing on his fallen body. His body may be depicted as in a state of sexual arousal. In her wild form, she appears as a horrible hag with a blue face, her hair sometimes containing snakes, wearing a necklace of human skulls. She wears children's corpses as earrings and cobras as bracelets; her purple lips stream with blood, her red tongue lolls from her mouth and she has sharp tusks for teeth. She has four arms which hold weapons and also make benevolent gestures, which shows her ambivalent nature.

In tantric worship, she is seen as both the terrible aspect of natural death and destruction, and the aspect of the life-giving mother. It was believed that she represented the wise truth that life and death were two sides of the same coin; one cannot exist without the other. Wisdom came from facing up to the terror and chaos of destruction and accepting and reconciling both of these opposites. Some texts show her wildness as the chaos out of which all creation occurs (including Brahma, Visnu and Siva), again harking back to the concept of the chaos of the original primaeval ocean.

Later Kali became worshipped more and more as the Great Mother, playing down her terrible aspects. As Kali Ma she is worshipped by many today as a benign and nurturing goddess.

Consider the opposites that are held within the tantric version of Kali: possessing a wild sexuality and also a terrible destructiveness. To embrace Kali means embracing the realities of life and death. Whereas we may be able to romanticise a warrior goddess, it is hard to romanticise a goddess who reminds us of little children struck down by disease or the accidents of childbirth. However in her wildness is also an incredible energy, a creative, life-giving and life-enhancing energy. Whereas civilisation likes to have everything neatly parcelled up, hiding away the opposites that we don't like, Kali represents the chaos of having everything to mind at once. And out of that chaos comes creation.

Ponder over your life, perhaps as a whole, perhaps as it is right now. Is there a chaos of good things and bad things? Do some of the aspects of your life drive you wild, perhaps the way things are changing, or failing to change, as you get older? How would it be to hold with the chaos for some time without trying to sort it out, and see what emerges? Possibly in terms of what happens, but perhaps more so in terms of what emerges for your thoughts and feelings?

Think of what your opposites are. For most of us life is definitely good and death definitely bad. For many in the Christian tradition, "purity" was good and sexuality bad. What are some of the opposites that you are aware of? Are you able to detach yourself from value judgements, and remind yourself when judging thoughts arise, to let them go? Sometimes we want to be all sweetness and light, and find it uncomfortable to say no to people or be firm about our boundaries. Think of how we owe it to ourselves to be able to feel the anger which alerts us to the fact that someone is taking liberties. On the other hand, you may be afraid of showing the vulnerable side of your nature that you hide within, and present to the world an image of someone who is totally unflappable and in control, or impervious to hurt. What contradictions can you find within your own nature? Can you allow yourself to be multifaceted, even contradictory, in honour of Kali?

Nammu

The Sumerians believed that the primeval ocean, personified by the goddess Nammu, had begotten alone a male sky and a female earth. Their son Enlil separated the sky from the earth and formed all living creatures. This original myth may have influenced the Hebrews who were taken into exile by the Babylonians, and who also referred in the

Bible to the world being created, by separation, out of the formless waters.

An (Anu in Akkadian, the later ruling power in the area), god of the sky, was the earliest known chief god in Sumer, connected to the city of Erech. He was the highest power in the universe, the king of the gods, but somewhat remote from human affairs. Later on the god of the city Nippur, Enlil, became the national god of Sumer. His name means Lord Air, and he chose the rulers of Sumer and Akkad. Enlil originally devised the laws that ruled the universe (the me's, that ended up in Inanna's possession). There were lesser astral deities such as the moon god Nanna, who controlled time, and the sun god Utu, the god of justice, and a number of humbler gods, spirits and demons.

The most prominent Goddess was Inanna. Ereshkigal was the goddess of the underworld and Ninhursag was Mother Earth and portrayed as a fertility goddess. There were also many lesser female deities.

Nammu, the goddess of the primeval sea, was possibly a former Great Goddess. Her formless waters of chaos held all the potential that gestated and then gave birth to earth and sky. She can inspire us to sit with the chaos of our terrors, confusions and tiredness, and not rush to try to find a solution.

Exercise: Nammu - Chaos

Find yourself a comfortable and quiet place and relax. You may want to tape this visualization so you can keep your eyes closed.

Imagine yourself as the first Creator Goddess, Nammu, Goddess of the primaeval sea, formlessness and chaos. Imagine a shimmering sky, cloudy white and grey and swirling and shot with flashes and edges of silver. Follow the sky down past a misty, blurred horizon until you find yourself looking at the primaeval sea, pearly grey and sparkling in the shimmering light. See how the small wavelets at the edge of the sea are breaking on wet silty sand where the vastness of the ocean meets the wide delta of a mighty river. Notice how the shimmering light reflecting off the pools and wetness of the mud makes this too of a pearly grey colour, reflecting the clouds and complementing the sea.

Now enter the sea, sink down into the formlessness of the grey waters. As it surges slowly past you, back and forth, there is no sense of direction, nothing to see above or below. Only the cold grey waters moving gently around you, cradling you. Now feel yourself merging with the waters, feel your becoming as one with the ocean. Imagine your limbs slowly dissolving into currents of water, your body formless and watery, filled with the primaeval wisdom and slow vastness of the first formless ocean.

Be aware of the chaos of this place, the female form of chaos. Nothing is defined, there are no words, no descriptions or naming or accounting. There is no control, only a vast letting go. Perhaps this mirrors aspects of your life: perhaps there is no sense of direction, the old certainties have gone, they have abandoned you or you have abandoned them. Imagine you are in a place of holding still, with nowhere to go and no need or hurry to find your sense of direction. Be content to just drift there in the primaeval waters; you are everything, so there is nowhere to go to, it is all you. How long have you felt like this: is it a short while; is it an eternity?

Now be aware of potential within your waters. Be aware that just as everything is encompassed by the infinite primaeval ocean of formless chaos, so also does the ocean hold within it all that can be. The ocean holds within it all potential; it is pregnant with all possibility. One day when the time is right, the ocean will mother forth all that needs to be, just as it should. You can have faith, you can trust that in your own good time, you the ocean will do what needs to be done, produce the action that needs to be taken. But for the present, this is only potential, there is no hurry.

Stay with the formless chaos, and note how this feels. Is it peaceful? restful? comforting? Do you feel the need, like the male gods, to rush around and do something, smash things up in an attempt to get things under control as soon as possible? Or can you hold with the chaos, allow things to gestate as long as they need to gestate?

Perhaps for you the chaos is unpleasant, an angry, swirling whirlpool that you wish to escape from. Perhaps possible solutions beat at you from all sides, to be discarded and reexamined as you twist and turn for solutions. Consider that the strength and degree and chaos is an indication of the creative forces churning within you. Hold still and let them churn, note their patterns and colours. Just watch and detach yourself. See how they appear, how you feel as they are part of you. Marvel at what energy and possibilities you hold with yourself. Feel the power of the Goddess.

Whether your gestating chaos be calm or restless, pleasant or uneasy, now imagine Nammu rising through the deep waters before you. Hear her reassure you that your gestation will last as long as it should, that you are pregnant with possibilities that need not be rushed. What else does she say to you? What do you want to ask her? What does she reply?

Now gently imagine yourself receding as a spectator of the waters; rising gently through them, seeing the sea and the shore and the sky. Now return to the room when you are ready, and start to move your limbs. When you are ready, open your eyes.

Draw what you remember of your visualisation, and then describe what you have drawn to a partner (or out loud to yourself, if alone).

Tiamat

As Sumer gave way to Akkad which in turn was conquered by Babylon, the male gods became stronger. Tiamat was the goddess representing chaos in the Old Babylonian creation myth, but after being slain by the god Marduk her body was used to create the sky and earth. The story is told in an old religious poem called the *enuma elish* – "When on high the heaven had not been named…"

Apsu was god of fresh waters, Tiamat goddess of salt waters, and Mummu was god of the clouds. There came forth Lahmu and Lahamu, representing the silt; then Anshar and Kishar, the twin horizons of sky and earth. These two begat Anu (An) who in turn begat Ea (Enki). We can see here the essential elements of southern Mesopotamia: where the fresh waters, necessary for life, carry the fertile silt and run down to the salt sea of the Persian Gulf. Georges Roux describes a misty morning at the present Iraqi sea shore, where the Shatt-el-Arab meets the sea, with mists and banks of cloud and "sea, sky and earth mixed in a nebulous, watery chaos".

Apsu and Tiamat then conceived many lesser deities, but they "troubled Tiamat's belly" and they decided to destroy them. The other gods were shocked, and Ea cast a spell on Mummu and paralysed him as well as putting Apsu to sleep and then slaying him. Ea then produced a son, Marduk, a great hero god, from his wife Damkina.

Tiamat had meanwhile declared war on the gods. She created dragons and serpents (often associated with goddesses) and placed her son, Kingu, at the head of an army. The gods were too frightened to fight. Marduk said he would take on Tiamat and her army on condition that he be made king of the gods. The gods agreed, and gave him the robes of a king, the sceptre and the throne.

Marduk armed himself with a bow and arrows, a mace, lightning, a net, and the four winds from each quarter. He raised up a storm calling Imhullu the hurricane and Abuba the tornado, riding them like a chariot, with a magic word on his lips and a healing plant in his palm. As soon as Tiamat's army saw him, they faltered and turned, and Kingu was captured. But Tiamat was undaunted. She roared defiance, gibbered and shrieked and shook and muttered spells.

They fought, and Marduk won. He caught Tiamat in his net, and as she opened her mouth he blew the four winds into her stomach. He split her belly with an arrow, smashed her skull with his mace and severed her arteries so her blood streamed in the wind. The gods laughed to see it.

Marduk now split apart Tiamat's body. With the upper half he made the arc of the sky, constructed so that the waters shall never escape. With the lower half he made the earth, and brought order to everything on it. Kingu's blood was used to fashion humans.

Compare this myth to the earlier, Sumerian creation myth. Note how Nammu, the primeval ocean, alone gave birth to the earth and sky. See how Nammu has become Tiamat and has now been demonised. Where once chaos was accepted as the mother of creation, now it is feared and must be destroyed first. Patriarchy in the person of Marduk blames her for all evil. Marduk has the male weapons of the arrow and the mace on his side, the storms and winds and lightning of the traditional patriarchal gods, the magic word on his lips – the word that categorises and names and controls, just as he orders and controls the setting out of the earth at the end of the myth when chaos has been vanquished.

Exercise: Tiamat - Anger

Find a quiet place where you won't be disturbed and relax, as above.

It may be that at this time of your life there is much to make you angry. Are you disrespected, ignored, passed over as an ageing woman? Are you angry at your insecurities – perhaps you fear what will happen when you can no longer earn a living? Would you like to strike out in new ways, but are too poor, too ill, or too dependent on a family you have spent years looking after. Perhaps you resent what you have spent your energy on over the years, be it a husband, or children, or a career which now does not offer you what you deserve.

You may be angry at the gods, at fate, for allowing you to age, and for being responsible for ageing in the first place. Perhaps you have physical problems, or emotional, or practical problems which make you feel life is unfair. Perhaps you simply have mood swings when you feel the anger rising within you, you are not sure whence or why. Whatever the cause may be, if you have anger within you, allow yourself to feel it now.

Having read the story of Tiamat and Marduk, imagine yourself at the scene of the battle between them. Imagine yourself as Tiamat, the Great Goddess, Lady of the Primaeval Ocean, Goddess of Chaos, holding within yourself the power and potentiality for the creation of life. See yourself thwarted by the upstart Marduk as he determines to control you, destroy you, impose his order and arrange things as he wants, determined to be king and ruler over everything, unable to bear the unknowingness, the as yet unrealised miracle of pregnant possibility.

Imagine yourself as Tiamat as she roars, raging, defiant, unafraid. Imagine yourself so angry that your wits are scattered, you shriek and cry out spells and maledictions, while your legs shake. Now imagine yourself with the might of the Great Goddess before she was dethroned by men. Imagine the fury of the storm whirling and raging around her, the four winds and the hurricane and the tornado. Imagine her commanding the winds to come to *her* command, imagine her pleasure as they roar and swirl and batter about her and she feels at one with them, knows that they are part of her chaos and part of her.

Grab the mace from Marduk, swat him aside with it like a fly. He is gone. Take his bow and quiver, fit the arrows to the bow one after the other and send them hurtling through the sky. Break the bow against your knee, tear the net to pieces and throw it all into the wind, smash the mace through the air again and again. Finally let go of the mace and send that hurling through the dark storm.

Feel the primaeval chaos of your anger and the stormy tempest all around you, let it blow and shriek and howl for as long as feels comfortable, imagine yourself shrieking and howling along with the wind, your arms raised into the wind, your body shaking in rage, your legs planted firmly apart, your head thrown back. In triumph and rage, be one with the tempest as you feel your defiance of all that the gods can throw at you, feel it from the top of your head to the tip of your toes. You are Tiamat, Goddess of Creation, Tiamat of the Primaeval Ocean, Tiamat the Mother of all worlds, Tiamat the mightiest and the first of all.

When the storm has raged for long enough, feel the winds gradually start to die down. Gradually the noise of the tempest, the howling and shrieking, begin to die down. Gradually all calms down, the wind is slow and gentle, the air is quiet. Tiamat has prevailed; her anger is spent.

Now Tiamat returns to her oceanic form. You are Tiamat, as she becomes the gentle waters under the sea. You know, as Tiamat, that you have power and strength, and when it is needed, you can get in touch with that power that you feel when you are angry.

Now you meet Tiamat face to face, in the deep ocean, this aspect of yourself. You tell Tiamat how you feel about the things that make you angry. What does Tiamat reply? You have a question for Tiamat - Tiamat replies. Is there anything else you need to say?

Tiamat wishes you well, and reminds you that you have both the strength to deal with what life throws at you, and the wisdom to know when to act as well as when to lie still as the ocean. She can act, but she reminds you that the ocean is also about potentiality, vast potential. As the formless ocean the outcome is not yet known, named, perceived. It is part of the greatness of the goddess that she can bear with not knowing, with just being, awaiting whatever may be. Tiamat now gives you a symbol of what you need, either as a talisman or as an indication of the next step to be taken, or of the time to act – whatever you need right now.

Gradually turn your imaging to the image of your own body, here in the room. Feel the extremities of your body, slightly move them. When you are ready, feel yourself back in the room and open you eyes. Get up and walk around a bit.

Now draw what you have visualised, and explain the drawing to a partner, or aloud to yourself if you don't have one.

CHAPTER 18: Emergency Self Help

There may be times when you just feel awful, so awful that you need some emotional first aid straight away and you haven't the energy to do anything particularly constructive. It would be helpful to work out some strategies for these times before they strike, but there are still things you can do to make yourself feel better even if you are unprepared.

If you are feeling terrible, or ill, you may need to work hard at keeping yourself on an even keel and not giving in to panic. This is not about achieving a better life, but simply getting to a place where you can calm down, let go of panic or despair, and comfort yourself as a mother might comfort a small child that is crying.

There may well be other things apart from the following list that you can think of to help you as emotional first aid. For you it might be going off for a long walk with the dog, or a good stiff drink. Anything that helps you immediately without ultimately making things worse (like too many good stiff drinks, or quarrelling with your partner) can go on your list for when you need it most.

Distraction

Sometimes you can ward off a bad time, or get through emotional or physical discomfort, simply by distracting yourself. This might mean going out, to the pictures or to see a friend, anything that sets your thoughts on a different line. If you're feeling really bad, you may be better off staying at home and distracting yourself. It depends what your interests are and what keeps you concentrated.

I find a book is the best way to distract myself. It has to be reasonably interesting but not requiring too much concentration. Best are old favourites which I haven't read lately, which I know are funny or have a happy ending.

For someone else, distracting themselves might involve writing a letter, or getting down to work, or writing a book, or gardening, or watching TV. Work out what it is for you, and if for example it is reading romantic novels, get in a good stock from the charity shops ready for emergencies.

Pampering

This may head off a fit of the blues. Again it might help if you made of list in advance of what constitutes 'pampering' for you. For me, it might be a nice long soak in a hot bath; eating something I know I'm not supposed to, especially chocolate; going out for retail therapy; watching an afternoon film on TV; basically doing something that would normally make me feel guilty. Sometimes you just have to accept that you can't face the world and you might as well enjoy yourself as you wait for the feeling to pass.

Supportive people

Think of people you know who can make you feel better, or even offer you practical help, according to the circumstances. I am lucky in that I have a very supportive husband, but if I want a long discussion about feelings he's out of his depth. I find my female friends can best help then. Make sure you're available when they need the favour returned!

Positive thoughts

We get into habits of thinking, and the more a thought occurs the more likely it is to recur again and again and the more difficult it is to counter it. Note down what your most common negative thoughts are that are causing you grief - in cognitive therapy these are sometimes called NATs (sounds like gnats) for Negative Automatic Thoughts.

Next, think of a counter to each negative thought – a positive thought (PAT) to put in its place. When you feel bad and the NATs are circling round your head, read out the positive thoughts to comfort yourself. The more you do it, the more easily you will remember the PATs and the less power the NATs will have over you. If you have difficulty thinking of anything positive, think what your mother might have said to comfort you – or if your mother wasn't a comforting sort of person (or wasn't very good at it!), think of what a friend might say. Or what you yourself would say to comfort a child, or a friend, or anyone in distress and who is having this particular negative thought.

Sometimes just writing down the negative thoughts can be enough to experience some sense of relief, because as soon as you see them written down they seem exaggerated and you automatically think things aren't so bad. But make sure you still write down the PATS as well. It doesn't matter if you are in such a state that you can't believe your PATS – by repeating them over and over to yourself, they will gradually sink in. Remember you probably don't realize just how much negativity your inner voice has been giving you. If you try to clear your mind of thoughts, you will soon see how much negative chatter there is. It is simply a matter of deliberately replacing the negative chatter with positive chatter.

Affirmations and talismen

Affirmations are very short sentences that should be repeated over and over to counteract a bad attack of the negatives. They are something to clutch onto when you feel you are drowning and are not really capable of anything constructive but distraction doesn't work.

Three weeks after my mother died I started my first job in a new career in social work. I was working with children in a local authority home and it was emotionally demanding. Sometimes at work in the first few weeks I would feel I might not be able to cope, and I would (silently) chant over and over to myself: I'm feeling OK, I'm doing fine. I'm feeling OK, I'm doing fine." Or it might be "I'm feeling fine, I'm doing OK. I'm feeling fine, I'm doing OK". This wasn't something to think about as to whether it was true or not, but simply a lifeline.

Closely allied was my own particular talisman, a small wooden cross that I held as I sat by my mother's bedside waiting for her to die from her stroke. I was just at the point of becoming a practicing Christian again and it had been significant to me for that reason, but now it also seemed to have become imbued with some kind of mothering symbolism. It was on a long chain round my neck and clutching it helped me to feel grounded. Apparently autistic children like to clutch something hard such as a toy car for a similar reason, and people throughout the ages have held onto lucky stones and suchlike.

If you are under a great deal of stress and strain for some reason, you may want to find your own affirmation and talisman. Perhaps you already have without realising it, and could make more use of them. A particular prayer might also help, whether or not you are religious.

You may well find various phrases from other books and articles and things people say that you can write down or mark to be gone over when attacked by negative thoughts. Each person must have the phrases that suit their personality. For example, many New Age type books have affirmations such as "I now feel well, energy flows through me" to say when you are feeling ill. It's not much use to me, but maybe it would work for you.

Inspirational sayings and images

You might be struck by something as you are reading, or listening to the radio; perhaps you may be given a picture or a quote or a poem on a retreat or transpersonal workshop, or come across something while on holiday or visiting a cathedral or mosque. A friend might send you a postcard or birthday card that particularly strikes you, or you may find something in a museum or art gallery.

There are also many books of inspirational sayings, perhaps on a spiritual theme, or words of comfort for dark times. The Bible is a rich source. The *Desiderata* by Max Ehrmann is a well known piece available in many book shops; I have found the following quote useful: "You are a child of the universe, no less than the trees and the stars; you have a right to be here. And whether or not it is clear to you, no doubt the universe is unfolding as it should."

For any kind of change that feels difficult, including the menopausal years: "One does not discover new lands without consenting to lose sight of the shore." (Andre Gide).

Based on Isaiah chapter 43: "Do not be afraid, for I have redeemed you; I have called you by your name; you are mine. When you walk through the waters, I'll be with you; you will never sink below the waves. When the fire is burning all around you, you will never be consumed by the flames. When the fear of loneliness is looming, then remember I am at your side. You are mine, O my child; I am your father (mother), and I love you with a perfect love."

During a difficult menopause I kept my PATs and sayings, references to particular books, pictures and postcards, etc in a folder, and wrote out quotes that I found particularly helpful in bad times in a little book. A good variety meant that even if one item failed to mean anything at a particular time, another was helpful. I have continued to find it useful since as ageing strikes me in different ways, and occasionally add to it.

Visualisations

In a visualisation, you get into a state of quiet relaxation and imagine yourself in a particular place or on a particular journey that you have prepared in advance or found in a book. Parts of it may be quite open ended and depend on your own imagination, or it may be completely worked out in advance so all you have to do is see it happening in your mind's eye. You might carry out a visualisation from this book.

It helps to be in a quiet place away from other people, where you know you won't be disturbed for half an hour or more. If you are in a bit of a state, you might want to help calm yourself first by listening to a tape of the sound of waves, or the wind through the grass (which you can buy from some bookshops, particularly New Age ones, or through the internet). When preparing or marking for future use a visualisation that is intended to help calm you down when you are ill or upset, it is best to have ones that you know will lead to a positive frame of mind.

Many books have visualisations in them, particularly New Age books on healing such as those by Gill Edwards (eg *Stepping Into the Magic*), or more medically oriented books such as Bernie Siegel's *Peace, Love and Healing* (he calls them meditations), or *Coping Successfully with Pain* by Neville Shone. *Getting Well Again* by O Carl Simonton, S Matthews-Simonton and JL Creighton is yet more specifically medical but has useful points on imagery. You may want simply to have the relevant pages marked and read through them quietly when required, stopping to give yourself time to close your eyes and properly imagine what they say.

A more effective method would be to read the visualisation onto a tape, taking plenty of time and leaving long pauses, so that you can just lie back and keep your eyes closed throughout. This would also give you the chance to amend the visualisation slightly to fit your exact circumstances or issues. Or you may be able to buy a tape prepared by an author, and a range of healing and calming visualisations can be found on the internet.

As well as longer visualisations, you can develop the habit of imaging a safe and quiet place to go to in your head, at any time. I imagine myself just off a golden beach in some warm, Mediterranean seaside place, floating in shallow water. The sun is shining, I feel deliciously warm but not too hot; the water is quite warm and feels silky as it laps at my skin. I am curled up almost in a foetal position, and the water is just the right depth so I am almost floating but also bumping slightly against the bottom. As the waves wash in with a slow swish, swish, I am gently moved back and forth. I have imagined this so often (I can't remember the first time) that it is relatively easy to go to that place now whenever I want to. I would guess it is so soothing because it somehow connects to old memories held within the body of being in the womb. I also have another scene, myself sitting in a sunny, flowery meadow that is an idealised picture of a place I used to go to years ago, and which is associated with happy memories.

Comforting tapes

Making your own tape for playing when you want to feel better can be much more effective than playing other people's, even if you borrow some of your stuff from books, internet and professional tapes. You might dip into your file of positive thoughts, read extracts from books or from other tapes, repeat your favourite quotations and prayers and inspirational thoughts.

You might make a tape about anything that bothers you, mental or emotional or physical. Just remember to be kind and caring to yourself, and remember to praise yourself for your many strengths and victories over adversity both now and in the past. If you can't think of any, ask a friend to help you make a list! If you didn't have any strengths at all you wouldn't be here.

If you have a specific problem, think how you would comfort a friend with such an issue and all the things you would say to make her feel safer, calmer, happier and more optimistic. Then when the fear arises in the middle of the night and there is no-one to talk to, you can play the tape and comfort yourself.

Books

You may find a whole book to be comforting, helpful or inspirational. You may also find that even though you don't like the entire book, parts of it are worth re-reading when you are feeling down. I find that some books are too positive and they don't help when I am feeling low; rather they inspire me when I am feeling ready for new projects or simply to feel even better.

I have gathered a number of books that I know are of use to me when I am really feeling down, and I keep them in a particular place. Thus I can easily lay my hands on them so that I can read them – or a particular chapter – when necessary.

I am very keen on books, so naturally turn to them. For you it may be something else – a particular painting, or film, or sculpture, or music, or park or garden.

Ritual

Sometimes if you are beyond the power of mere words to reach you, some sort of ritual can be very soothing. People find that even if they have left the religion of their childhood behind, to go into a cathedral for example and light a candle or sit quietly for a while can be a calming experience. Repeating an old prayer may help. Attending a church service, taking part in the Eucharist, even if you are not paying attention to the actual words, soaks in at a deeper level.

You may make up your own ritual. One example might be a ritual washing away of what is bothering you – stepping into the shower and letting the warm water wash away the unpleasantness of some other person's meanness, or the burden of your various duties to others, or worries about the future. Or you can sit in a special corner of your home which perhaps holds some treasured objects of significance to you, where you light a candle to renewed hope, or peace, or the future, or absent friends.

Or you might try an incantation from a book such as *Good Magic* by Marina Medici, perhaps first running a warm bath with some aromatherapy oil in it and lighting the room with candles.

I do not believe in magic literally, but I do believe in the power of ritual and symbol to affect our emotions and sense of self. Again, you will find all sorts of incantations and rituals – particularly goddess-related – on the internet.

Emergency self counselling

There are a number of methods of counselling that lend themselves to emotional first aid. The business of positive thoughts is an aspect of cognitive therapy, for example. I will just give three very simplified examples based on transactional analysis, gestalt therapy and solution-focussed brief counselling.

Transactional analysis (originally developed by Eric Berne) suggests that we find ourselves in different 'ego states' at different times, relating to how we are feeling and behaving. We may be in Child state, which tends to be about emotion – either the free Child, enjoying herself and playing, or the victim Child, being told off or persecuted by a Parent figure. In the Parent state, we are of course acting like a parent – either nurturing (comforting or managing and interfering) or persecuting (telling off, saying things like "you always…." "you never….."). In the Adult state, we are more purely ourselves, without slipping into role playing, able to be clear sighted and logical and emotionally on an even keel.

If you can recognise which state you are in, Child, Adult or Parent, that can go some way to putting yourself back into Adult. Just the intellectual work and detachment necessary to think about it can be enough. If you are feeling sorry for yourself, that's Child; feeling guilty is your Parent berating your own Child; unable to shake off a righteous anger, thinking of all the things you should have said if only you'd thought of them, means you're in Parent. We usually get into habits of these ego states while we're literally a child, and often by learning them off our elders.

Deciding to do something about it can put you back into Adult. It can also help when feeling bad, to think: "When have I felt like this before? What was going on?" You may be able to trace certain habits back to an earlier time, to a deeply felt response which has now been triggered by a similar episode. When your Adult has remembered why this hurt so much in the past, you should be able to be more philosophical about it now you are grown up. At least you might recognize that you now have a choice as to how to respond – you can choose to work at not feeling hurt, scared, or victimised.

One Gestalt method is to role play different parts of yourself. You decide in advance which parts you will play, and allot a different seat for each part. The permutations are endless, according to what's going on and what you want to learn, to integrate or otherwise achieve. For example, if you are upset and you don't know why, you may decree three parts: the normal you, the Wise Woman you, and the upset you. By moving from chair to chair, you can ask the upset part of you what is going on, how she feels and why, and if she can't identify anything exactly, then how it feels to be upset which helps to pin it down.

For self comfort, you may again decree three aspects of yourself: the normal you, the upset you and the Loving Friend you. Then you can have a dialogue between the upset you and the Loving Friend, letting the upset you explain how she feels, what her fears are or whatever, and the Loving Friend will respond with words of comfort.

Remember when doing this sort of thing to end up back in the chair as the normal you, and make sure you are properly grounded again by stamping your feet on the ground, getting up and stretching and perhaps brushing your hair. Always include a wise or comforting part of you. Don't work with anything too challenging unless you are with a gestalt therapist, or you might simply get more upset.

For the solution-focussed therapy, when you are feeling bad about yourself or a problem, draw a diagram of your life and all the different parts that go to make it up. That might include family and friends, work whether paid or voluntary, interests, hobbies and enthusiasms, places you go and people you meet, organisations you are part of, plans and ambitions, things that inspire you, beliefs and experiences that are significant to you and which you feel go to make you who you are. The purpose of this is to show how your feelings or the problem are just one part of who you are, and not necessarily the most important part. Even if you have terminal cancer, there will be people, philosophies and ethics, activities and interest that are important to you, and experiences and achievements to look back on and future goals to be met. You should also reflect on what qualities you have shown to get you so far.

Next, think of what life would look like if you were feeling OK in your emotions. This does not necessarily mean that your problems are solved, simply that they are not making you feel so distressed – perhaps because you are coping better. Think about how you would be feeling, what you would be doing, who would be around you, how you would be relating to people etc. Now think of all the ways in which you are already like that or have been in that space already. What qualities did you use in order to be there, which are at your disposal to draw on in the future?

One example would be feeling low because your opportunities have been curtailed. Perhaps you feel realistically that you will never make it in your career now because you are too old. Think of other times when you have faced disappointments and come through, and have managed to accept and detach and move on to other paths. How did you have the strength to do that? What does it say about you?

Alternatively, take some time to imagine yourself in a future place when you feel fine about it all. You remember how bad you used to feel and want to give yourself advice now. Where are you? What are you doing? What are you thinking about your old problems? How did you get to this point? What advice do you give to your old self, what is the next step to take? What resources did you use – which you have right now – to help you to get to this calm place?

Of course, it is always better to have counselling from a counsellor since the relationship is part, some would say the most important part, of the healing process, but not everyone can access it and it may take time to set it up. Self-counselling may help in an emergency, and it doesn't cost anything. Once again you can also find counselling self-help on the internet, particularly cognitive therapy (CT) and a tapping technique based on CT and acupuncture called the Emotional Freedom Technique.

Artwork

Sometimes self expression can release an emotional blockage and enable conflicted feelings to calm down. Just drawing whatever picture comes into your head may make you feel better. At a time of anxiety I tended to draw pictures of myself in a boat on the ocean with sharks circling round. Another recurring theme was myself at the bottom of a dark cellar or underground. Anger might be a volcano erupting, fear of a new project might produce a picture of canoeing over rapids. Somehow, just letting the images emerge on to paper released something within me.

For you, producing a poem about how you feel might help, or strumming on a guitar, or cutting up some magazines and producing a collage. Letting yourself go with a big lump of wet clay is very soothing as you work it with your hands. Perhaps working in the garden would have a similar effect.

Tarot

I don't believe in the power of the Tarot to tell fortunes, but my friends and I do enjoy telling each other's fortunes with the cards all the same. As well as being harmless fun (but first remove the three of swords, which can foretell future misery, just in case it turns up and sticks in your mind!), it can be another self-counselling tool. You can interpret the cards to yourself - their basic meaning is given in the accompanying booklet to the cards, or you may have a specialised book, but you have to work out how to apply it to your own life now. It helps you to sort out what your feelings are about yourself and your situation, as well as your hopes and fears for the future.

Dreamwork

Sometimes when you are upset about something you have a dream that sticks in your mind. Probably most of our dreams are of no psychic import, and that's why we forget them. I believe that if one sticks in your mind, it probably has a message for you. It is a good idea to write it down straight away so you don't forget the details.

One way of interpreting dreams is to tell the dream in the first person, in the present, for each person or thing that appears in your dream. Since only your psyche produced the dream, each part of it must represent part of you. If you dreamed of your mother turning up, she wasn't actually there; thus, her figure in your dream represents a part of you that feels like some aspect of your mother.

To give an example, I dreamed of being in a musty old church hall, where there were a number of elderly people. Some of us were dancing, but I felt pretty awkward and not sure I wanted to be there at all. I didn't feel part of it. There was an old chest in a corner, which I opened to find some crumbly old biscuits and some old pennies (pre-decimalisation). The biscuits and pennies started to spill out so I hastily closed the lid. Again there was the feeling that I didn't really want to be there.

When I went over the dream later in the first person, telling the story in the present, what struck me was the emphasis on "old" – everything was old. Could this be about old age? It was a very negative view of oldness – musty, depressing, rather lonely, a strong feeling that I didn't really want to be here and didn't really belong to this club!

When I told the dream from the point of view of the chest, I described myself as "an old bag". The biscuits were too old to be nourishing, nobody would find me appealing anyway. As the coins, I was out of date and again no use to anyone. The chest being opened and the contents spilling out, myself wanting to push them back in – did this represent my ambivalent feelings about ageing consciously, almost wallowing in thoughts of old age and approaching death through writing this book and giving workshops? I realised part of me would far rather just forget the whole thing and ignore it, to try and make it less real.

Sometimes we feel bad about something just because that is the appropriate response; grieving the loss of something or someone important, for example. Other times we feel bad because there is conflict within us. Perhaps we feel something but don't want to acknowledge it to ourselves. This dream was about conflict between two parts of me: the part that believes the unexamined life is not worth living, and the part which thinks eat, drink and be merry - and don't remember that tomorrow we die.

Count your blessings

Make a list of all the things that you have to be grateful for. This may be aspects of the world that we all share, such as blue skies and beautiful flowers; people you love or have loved; experiences you have found interesting or enjoyable; everything you enjoy about life or would miss if you didn't have it. You may also include what you have gained even while you have been facing adversity.

Remember the words from Kahlil Gibran (*The Prophet*), about joy and sorrow: "Together they come, and when one sits alone with you at your board, remember that the other is asleep upon your bed." Just as none of us can avoid dark times in our lives, neither should the good times ever be forgotten.

CHAPTER 19 - The Path of the Elderwoman

We have looked at how women in the last third of their life are burdened by social stereotypes which are less than helpful in giving them the sense of power, strength and wisdom needed to face the challenges ahead. We saw how we need to redefine how we see ourselves and how we find meaning in our lives. Archetypes that can help us to do that include those of the Hero, the Goddess and the Wise Woman. Together they make us the Elderwoman, the female counterpart to the Alderman (from the word for Saxon tribal elders).

In this chapter we look at where we might be going with our lives, what our next steps might be, once we have fortified ourselves with new images of our courage and self esteem. We start with a couple of exercises to help focus our minds on any unfinished business we may need to attend to, and then explore what particular path we as individuals might be following.

Exercise 1: Looking back.

Find a quiet uninterrupted space and relax. Now imagine yourself on your death bed. Take some time to see yourself lying in the bed, see the details of the size of the bed, the coverings, the time of day. There has been plenty of time for family and friends to gather; this is a typical Victorian death bed romance. Although you are weak you can easily talk and register what is going on.

Who do you imagine is there at your bedside? What are their emotions as they see you lying there? What are your feelings about them? Is there anyone missing, whom you wish could come? As you look at the people around you and think of your relationships, is there any sense of unfinished business here – perhaps words never spoken that should have been, words of love and praise and gratitude that were held back and now you wish had been spoken out loud? Or are there any words you wish unspoken, perhaps an ancient quarrel or misunderstanding that was never resolved? Do the people who matter to you know that you care?

Now cast your mind over more practical matters. Have you left your affairs in good order? Have you made provision for those that need it? And what about your life as a whole – have you done those things you wanted to do, do you have any regrets about the way your life has gone? If so, is there anything in particular you wish you had done?

Now come back into the present, ground yourself by walking round the room, and reassure yourself that you are still going strong! If this little exercise produced any sense of things needing doing, jot them down. Consider how realistic it is to do something about them - and remember it's highly unlikely that anyone can get through life without having any regrets at all. But there may be one or two items needing to be resolved.

Exercise 2: Your obituary.
A similar exercise you might try is that of writing your obituary, once by imagining how others might write it if you died today, and a second time as you would like it to be written. You may well be fortunate enough to find that the two are the same.

Exercise 3: A life audit.
Draw a circle in the middle of a large sheet of paper, and write "My Life" in it. Now draw pathways radiating out, with different colours for each main pathway. Each pathway may branch into other pathways. Name the main pathways after the main areas of your life. Some examples are: work, social, family, hobbies, education, church, political party, voluntary work, alone time, nature, spirituality, money, physical or mental health, leisure, service to others, home, creativity. Smaller branches arising from "alone time", for example, might be named reading, walking, sitting in the garden, diary. You may want to draw little pictures or symbols against some of the items. Write a word or two for each activity, which indicates how you feel about it (for example: rewarding; not enough time; boring; feel obligated; very important).

Now reflect on your drawing. In which areas do you feel fulfilled, alive, challenged? Or peaceful, relaxed, renewed, refreshed? Or drained, trapped, frustrated? Would you like to develop more branches, or more sub-branches for a particular area? Would you like to prune them? Improve them? What do you really like doing? What must you do? Are there things you would like to do that aren't down there? You may want to make notes.

If you feel a sense of frustration for your life as it is now, you have to decide whether this is the time to do something about it, or time to let things lie until the way forward becomes clear to you. The seasons must come round in their own time and cannot be hurried, but equally a task that is due now should not be put off. Part of wisdom is knowing when to act.

The following exercises may help you to consider which path you wish to follow now and what the next steps, if any, might be. It may well be that you are reasonably content with your life as a whole, but wish to develop or concentrate on just one particular area; or you may be wondering about the whole shape of your life.

Exercise 4: Visualising your future path.

Get yourself comfortable in your chair, or lying down, at a time when you won't be disturbed. You may want to record this and play it back with your eyes closed – remember to leave plenty of spaces to allow time for your imagination to work. Have drawing materials ready by your side for when you have finished. You can do this visualisation more than once, as it will change as your life continues to unfold.

Imagine you are standing on a path. The sun is shining, it's a warm day but not too hot. What does the path look like? What is around you? What are you wearing? Now imagine you are travelling along the path in front of you. Keep aware as you move along, what is around you, ahead of you. How do you feel as you go along, how fast are you going? Is anyone with you? Are you taking anything with you?

A little further ahead there is a rise in the path; once over that hill you will see your destination. Just before the hill is an obstacle. See the obstacle and what it consists of. How do you get past it? Now you are past the obstacle. Was it easy or hard? What helped you? Once past the obstacle you are at the brow of the hill and can see your destination. Note what you can see at this point.

Go down the hill to your destination, and note what you can see around you. What meaning does this have for you, if any? What emotions do you feel as you stand there? What would you do next? What needs to be done, if anything, or what do you want to do? What plans do you have, if any? How do you feel about it?

You now see a figure approaching you. This is your Inner Guide. Take time to note the appearance of your Guide and to feel at ease with him/her/it. Now you can ask your Guide any questions you like about what you see around you, and listen to their answers. When you have asked all you want to ask, your Guide gives you something that symbolises this destination and what it means to you. In a moment you will be leaving this place.

Your Guide now takes you back to the place where you started your journey. Take a few moments to reorientate yourself. As you look around you, your Guide points out a narrow path that you didn't notice earlier, that leads off to the side, away from the main path. Your Guide now urges you to follow this new pathway, and follows behind you. Is there anything you need to take with you on this different pathway? You set off – note how the path looks as you go along, what is around you. Is there any obstacle on this path? If so, how do you overcome it? Keep following the path. How easy is it? How are you feeling? How fast are you going?

Now you are reaching your alternative destination. Take some time to let it become clear before your eyes. Where have you come to? What does it look like? What can you see? How do you feel about this place/? What would you do next? What do you expect? If you have any questions about this alternative, ask your Guide. When you have finished, once again your Guide gives you something, which symbolises what this destination means to you.

Now follow the Guide back to your starting place. Say farewell to your guide, but know that you can meet your guide again whenever you wish, simply by going within. Turn your attention back to your body, feel yourself becoming aware of the room again. Gently start moving your limbs and open your eyes.

Now draw what you can remember of your journey, remembering that the purpose is to make a record and to draw out new meanings, not to produce a beautiful work of art (though you may do so!) You may want to make one or two notes, but try to keep writing to a minimum. When you have finished, if you are able, describe what you have drawn to a partner – sometimes we find a particular meaning and significance only as we hear ourselves speak. If you have no partner, speak out loud to yourself. See if you can relate what you have experienced to real issues in your current life, and the choices of path before you.

Exercise 5: The next step.

Get yourself in a comfortable position where you won't be disturbed. In this visualisation we are going to call on Hecate's help. You will need three cushions. Remember during this visualisation that although its purpose is to help you visualise your next steps in life on the path towards your desired destination, this may be a time when you need to lie fallow and wait for the right season; your next step may be to accept that need.

Imagine yourself walking along a path towards a vast cave; the time is dusk. You enter the cave, to find it ablaze with candles in nooks around the walls. The temperature is just comfortable for you, and the ground beneath your feet is soft clean sand. Take some time to feel at home within the cave.

Now Hecate appears, with three cushions. As she indicates that you should move from one cushion to another, do so in real life, on the cushions that you have set out on the floor around you. The first cushion she asks you to sit on is the seat of the material world, of work, career, money, ambition, plans and projects. Sit on your cushion. Take some time to gently run your mind over the concepts of the material world, and let images spontaneously arise – watch your images arise in as detached a manner as possible, rather than getting involved in them.

Now Hecate asks you what you want from this material world; what comes to mind? Do not think about this logically if you can avoid it, but simply let images come to you if they will. If nothing comes, don't force it. Now ask Hecate what your next step should be; listen carefully to her reply. You may want to ask Hecate other questions while you sit on this cushion.

Now move to the second cushion, both in the cave of your imagination and in real life. This is the cushion of emotion, relationship, your social life; of relationships with family, friends, colleagues and others. As before, let images gently arise in your mind, without forcing it. Hear Hecate asking you what you want from this world of relationship, and see what arises in your mind's eye. Ask Hecate what your next step should be; note her reply.

Now move to the third cushion, the cushion of your Self, of the personal, the spiritual, of interests, of what gives you beauty, comfort, pleasure, meaning, a sense of fulfilment, what intuitively feels right. Take time to let the images of the third cushion rise before you. Hear Hecate asking what you want for this inner world, and see what images arise for you. Ask Hecate what your next step should be; note her reply. As before, you may ask her further questions if you need to.

Now return to your original seat. Thank Hecate for her advice and wisdom. Ask her any further questions that you have; you may wish to be shown some sign or symbol of what your next steps should be. When you are ready, leave the cave and make ready to return to the room. Now open your eyes. Again, you may gain more insight by drawing what you have imagined.

Aspects of Elderwoman

Here we shall have a look at some of the variety of types to be found in the old myths of the Goddess. Read through them and think about which Goddess resonates most for you. Which Goddess are you most like? Which Goddess inspires you? Which qualities would you most like to develop?

The Wyrd Sisters – power of life and death

The Wyrd Sisters, or the Three Sisters of Anglo-Saxon England in the first millennium were common to much of Western Europe: called the Norns in Scandinavia, the Parcae in Rome and the Moirai in Greece. They were in charge of the threads of life; they held the power of life and death.

The three sisters spin the web of life, and everyone's own particular thread or life line contributes to their tapestry. When a child is born its thread becomes part of the tapestry; when it finally dies, young or old, the thread is cut. Thus all are interconnected. What I like about this imagery is that a life remains in the tapestry even after death. The part played by a living person lives on, and the effect they have had, the interrelationships, the colour and thickness of thread, the contribution to the overall pattern – it all remains.

Spinning and weaving, as any other art or craft, had its sacred aspects, and the wonder of it was not taken for granted for a very long time. There was usually a patron goddess of spinning and weaving, or weaving might be used as a metaphor for divine creation. Arachne in Greece was turned into a spider by Athena after she dared to challenge her to a weaving contest.

The three sisters were the spinners of creation. When a child was due to be born, the sisters gathered at the house. Brian Bates in *The Wisdom of the Wyrd* reports that this was still thought to happen as late as the eleventh century in Germany, and the Christian Bishop of Worms complained that women would lay three places at the table for the Parcae. In the case of the Moirai, one sister would do the spinning (Clotho – Nona among Romans); another would draw lots to determine the length of the thread (Lachesis/Decuma), and the third would cut the thread at the appropriate length (Atropus/Morta). The Wyrd sisters probably did the same.

In Iceland the sisters were called Urdr meaning life unfolding; Verdandi, becoming; and Skuld, which has a meaning of a debt to be repaid. We have been given life, and we must pay it back in the end.

I recommend Brian Bates' book for an account of the way of Wyrd and in particular a very interesting discussion of the meaning of the web of life. Here I would simply point out the power of these sisters: to determine the fate of each human being, the course of their life and how they relate to each other. And it was female deities that wielded that power and were respected accordingly. Do you feel that you have enough power in your own life? Have you given too much away? Are there any ways in which you can reclaim some power and respect? If it is not immediately clear, could banding together with others in the same position give you some power as a group?

Sekhmet – the courage of a lion

Sekhmet was originally the war goddess of Upper Egypt. She was depicted as a lioness, or a woman with the head of a lion. Tame lions were kept in her temple. She protected the pharaoh in battle, killing his enemies with fiery arrows. Her breath was believed to be the hot desert winds.

At the end of battle, to pacify Sekhmet, festivals were held to make sure that the destruction would cease (this may have been a symbolic way of calming down the bloodlust in those who had been fighting). People danced and played music to soothe the Goddess, and drank a great deal of beer.

There was a myth that Ra created Sekhmet to destroy his enemies who were plotting against him, but having dispatched his enemies she went on to kill more and more people, crazed with bloodlust. Eventually Ra tricked her by colouring drugged beer red with ochre so she thought it was blood and drank down great quantities, making her so drunk that she stopped fighting and became calm again. As Sekhmet was full of bloodlust, she was identified with the colour red and with menstruation.

Sekhmet was not the instigator of violence, but protected the pharaoh and his people when called upon. She was also goddess both of disease and healing, and her priests often acted as physicians.

Sekhmet reminds us that at one time, female deities were seen as having the attributes of courage and strength sufficient to fight and win battles. Many other warrior goddesses exist, such as the Morrigan (Celts), Anat (Canaan), Inanna (Sumer), Athena (Greece). Do you feel that your courage needs developing? Do you need to reach within you to find an inner courage? Visualisations and affirmations can help you get in touch with your courage if you feel you have lost it; so does reviewing your life and remembering all those times that you too showed courage. You are unlikely to have fought in battle, but you will have fought your own private battles against shyness, low self-esteem, the desire to give in. No-one can get through life without displaying courage at some points, so don't be modest.

Cerridwen - wisdom

Cerridwen was a Welsh (Celtic) nature goddess of wisdom, poetry, the moon and grain whose symbol, the white sow, represented the moon. She was known as the White Lady of Inspiration and Death. She was often depicted as a crone or mountain mother who let fall huge stones from her apron (or womb) which later became mountains.

She was mother of two children, a girl Creirwy who was beautiful, and a boy Afagddu who was very ugly. Cerridwen had a large cauldron in which she brewed magic potions, and she decided to brew a potion of inspiration that would make Afagddu all-wise. This potion was called greal and was made from six herbs for inspiration and knowledge. A youth named Gwion was set to stirring the mixture, but when she was out three drops of the mixture splashed onto Gwion's hand and he immediately put his hand to his mouth and sucked it. He thus drank all the wisdom of the cauldron.

When Cerridwen returned and found out what he had done, Cerridwen went for Gwion who ran away, now able to shape shift into various animal forms. Eventually Cerridwen caught and ate him while she was in the form of a hen, but later she gave birth to him in the form of Taliesin, the Welsh bard.

How easily can you identify with Cerridwen? She is more like us in her activities, a creative, nurturing housewife, cooking, caring for her children. When in need of inspiration or wisdom, perhaps we can visualise ourselves drinking from Cerridwen's Cauldron; identifying with her, we can make contact with the wisdom and creativity that lie within us, if we allow ourselves to be patient and let it gently rise to our consciousness.

Kuan Yin - compassion

Originally Kuan Yin was male. In Buddhism a Bodhisattva is one who has attained enlightenment, but instead of passing into nirvana decides to stay on earth out of compassion for the suffering creatures still bound there. Avalokitesvara was such a bodhisattva in Sanskrit writings, and eventually his veneration passed into China in about the first century CE, also reaching Korea and Japan.

In China he became known as Kuan Yin. He was considered the personification of compassion, kindness and wisdom, qualities traditionally associated with a mother goddess. Gradually Kuan Yin came to be seen as female, and depicted as a beautiful woman in white. By the 11th or 12th century nearly all statues show her as female. According to legend, she was granted many heads with which to hear the cries of the suffering, and many arms with which to help them.

Kuan Yin had risen through many incarnations to be a person of wisdom and service, and was on the brink of nirvana when she heard cries of woe coming from the many creatures of the world. Moved to compassion, she resolved to return until all had preceded her into nirvana. Another legend tells of how she begged her father to be allowed to enter a temple instead of marrying; he allowed her to do so but ordered her to be worked very hard to discourage her. She was so good that the animals helped; her father then tried to burn down the temple. Kuan Yin put out the fire with her bare hands without being burned, and her father out of fear ordered her killed. Because she was so good and kind she was made into a goddess and was journeying to heaven, when she heard the cries of suffering from earth. She asked to be sent back as a bodhisattva.

Kuan Yin teaches that the more we are awakened to our true spiritual nature, the more we feel compassion for others. When our minds are still and clear (for example through some practice such as yoga) then we realise that all earthly matter is impermanent, and only the divine within us is eternal. Attachment to the illusions of this earth is what causes suffering. Once able to feel detached from this earthly body, we lose our fear of death and suffering. Once awakened to our true nature, full of compassion, we become Kuan Yin.

As we attain wisdom and knowledge with age, we can use it as far as possible to help others who are not so wise, rather than becoming impatient or angry with them. When others give us a hard time – annoying adolescents, for example – we can remember how difficult it is to be young, how unaware young people are of their own feelings and motivations, how desperately important things seem to them such as fitting in with their peer group, and how overwhelmed by their emotions they can be. We in our turn can sometimes allow ourselves to be caught up in the drama of it all. It may help to imagine ourselves as a lotus flower, bobbing along on the sea of life: storms may rage around us (even in our own heads), the water may be agitated into waves beneath us, but we remain above it, bobbing serenely along.

With a sense of detachment, perhaps seeing everyone, including ourselves, as players acting out roles upon a stage, we are better able to see how little it matters in the long run, since in the long run we have to leave this life anyway. What does matter is caring and compassion while we are here. Compassion is traditionally a female quality, and Kuan Yin has often been compared to the Virgin Mary. As women we might feel that it ought to be easy for us to feel compassion, but find instead that we identify more closely with Sekhmet! When we need compassion, it might help us to remember Kuan Yin.

Hathor – love and joy

In Egypt, cattle herding began about 7,000 BCE in the Sahara and on the west bank of the Nile. The cow was seen as a source of life and nurture and thus several cow goddesses arose including Hathor. After severe droughts in 6,000-5,000 BCE the people settled on the banks of the Nile, farming wheat, barley and domesticated cows.

As the local chiefdoms amalgamated to become agrarian village communities from about 4,500 BCE, local gods and goddesses gradually acquired wider influence. Egypt became a single nation state in about 3,200 BCE with a national king, and myths were used to validate his position. Thus he became a descendent of a god king, with female goddesses becoming less important. Pharaohs became the manifestation of the god Horus, depicted as a falcon.

Hathor became a major goddess and was originally the goddess of the Milky Way, which was supposed to be milk scattered across the heavens from the udders of a heavenly cow. Hathor was depicted as a cow or given a human face but with cow ears, and as the nurturing cow was goddess of motherhood. She was the mother of the god Horus and thus became the divine mother of the pharaoh; she therefore became wife of the sun god Ra.

The sun god sailed along the Milky Way, which was called the Nile in the sky. By association, Hathor was seen as responsible for the yearly flooding of the Nile, which fertilised and irrigated the land. This confirmed her in her role as goddess of fertility, and being associated with the beauty of the nourishing cow, she became a patron of lovers and sexuality.

On one occasion the sun god Ra was upset about the victory of Horus over the god Set, who represented respectively Upper Egypt and Lower Egypt, and Ra went off into seclusion (it has been suggested that this symbolises an eclipse of the sun). Hathor went to him and started to dance, stripping naked and showing her vulva in order to cheer him up.

Hathor became amalgamated with another goddess who was connected to the soul, and thus became the greeter of the souls of the dead in the afterlife, giving them food and drink. This other goddess, Bata, was also associated with a musical instrument called the sistrum and Hathor became associated with music. She was celebrated in this form at her temple in Dendera, which had columns carved in the shapes of sistrums, and her priests were also dancers and singers.

Hathor was now protector mother goddess of all women, goddess of sexuality and fertility and lovers, especially kind protector of the newly dead, and goddess of jubilation, dance, music, and inebriation. One of her titles was "Lady of the House of Jubilation", another "The One who Fills the Sanctuary with Joy". She was one of the most popular deities in Egypt, particularly among women.

In later years the amalgamation of the deities continued. Eventually Hathor became merged with Isis, whose cult was also very popular and widespread.

Hathor seems to have shown so many different kinds of love – for son, husband, lovers, women, the poor souls entering the underworld. She was gentle and kind, also bawdy and joyous and noisy at times. That kind of generous love of people, land and life, overflowing like milk, is the kind of love we could all aspire to.

Exercise 6: Goddess visualisation.
You may want to visualise receiving those qualities from the various goddesses mentioned here, or perhaps you have favourites of your own that you would like to add. Having got yourself comfortable, imagine yourself in a beautiful, holy space, perhaps some ancient temple with limestone pillars and marble floor, open to sky and the green countryside. As you sit on a velvet chair, one by one the goddesses come to you, bearing a symbol of their qualities: the three Wyrd sisters, carrying their weaving equipment, to give you a spindle of power; Sekhmet with her lion head, giving you a lock of her coarse lion's hair to symbolise courage; Cerridwen with a small phial of liquid from her magic cauldron, symbolising wisdom and inspiration; Kuan Yin robed in white, with a lotus flower symbolising compassion; and Hathor with her cow ears, with a tiny cup of milk to symbolise love and joy.

It may be that none of the goddesses we have looked at in this book are quite right for you. Remember that there are goddesses from all over the world and from many different eras. A search through libraries, bookshops or the internet is sure to turn up one who captures those qualities which you are searching for; it may even be her image that you are seeking for. A selection of books is shown in the Bibliography, and I particularly recommend *Goddesses for Older Women* by Jean Shinola Boden.

Honouring Elderwoman
Let us complete our journey by taking a moment to honour ourselves and our sisters and to wish us all well on our journey through the last third of life. Find a peaceful space and light a candle to Elderwomen everywhere.

In a final visualisation, feel part of the sisterhood of women who are ageing or going through the menopause. Imagine gathering with other women in your position, see what they look like, how you greet each other. You are not alone. Imagine the compassion that we all have each for the other as we become aware of each other's pain, and rejoice in each other's happiness. Imagine us all linking arms to face the future together. Think of what we might say to each other to support each other and to honour women like us all over the world. See how all our faces shine as reflections of the Goddess, and how we are held in her embrace. Remember that the journey we are undertaking has been undertaken by women for aeons and sanctified by their courage, wisdom and laughter. Together we are strong.

Bibliography

Chapter 1

Adams-Price, C., Goodman, M., Oppenheimer, B., Codling, J., Roberts, P., Albrecht, B.C. *An Analysis of Aging Women in Film and Television.* Mississippi State University. http://130.18.140.19/conference/index.html

Austen, Jane (1818). *Northanger Abbey*. (2003 edition) London: Penguin Classics.

De Beauvoir, Simone (1949). *The Second Sex.* (1972 edition) New York: Penguin.

De Beauvoir, Simone (1970). *The Coming of Age.* (1996 edition) London: Norton.

Dillaway, Heather (2005). Changing menopausal bodies: how women think and act in the face of a reproductive transition and gendered beauty ideals. *Sex Roles: A Journal of Research*, July 2005.

Fox, Kate (1997). *Mirror, Mirror: a summary of research findings on body image.* Social Issues Research Centre (www.sirc.org/public/mirror.html).

Greer, Germaine (1991). *The Change.* London: Hamish Hamilton.

Greer, Germaine (1999). *The Whole Woman.* London: Transworld Publishers.

Schneir, Miriam (ed) (1995). *The Vintage Book of Feminism.* London: Vintage.

Wolf, Naomi (1990). *The Beauty Myth.* London: Vintage.

Chapter 2

Douglas, Claire (1990). *The Woman in the Mirror: Analytical Psychology and the Feminine.* Boston: Sigo Press.

Edinger, E.F. (1992). *Ego and Archetype.* London: Shambhala.

Fordham, Frieda (1973). *An Introduction to Jung's Psychology.* Harmondsworth: Penguin (3rd edn).

Hall, Nor (1980). *The Moon and the Virgin: Reflections on the Archetypal Feminine.* London: The Women's Press Ltd.

Jung, Carl G. (1968). *The Archetypes and the Collective Unconscious.* London: Routledge (2nd edn).

Jung, Carl G. (1964). *Man and His Symbols.* (1978 edition) London: Pan Books.

Jung, Carl G. (1953). *Four Archetypes.* (2003 edition) London: Routledge Classics.
Pascal, E. (1994). *Jung to Live By.* London: Souvenir Press.
Pearson, Carol (1991). *Awakening the Heroes Within.* New York: Harper Collins.

Chapter 3

Bettelheim, B. (1976). *The Uses of Enchantment: the Meaning and Importance of Fairy Tales.* (1991 edition) London: Penguin.
Campbell, Joseph (1949). *The Hero With a Thousand Faces.* (1993 edition) London: Fontana Press.
Maslow, A.H. (1970). *Motivation and Personality.* New York: Harper and Row.
Maslow, A.H. (1971). *The Farther Reaches of Human Nature.* New York: Viking Press.
Rowan, John. (1993) *The Transpersonal.* London: Routledge.
Somers, B. with Gordon-Brown I., ed. Hazel Marshall (2002). *Journey in Depth: A Transpersonal Prespective.* Cropston: Archive Publishing.
Vogler, Christopher (2007). *The Writer's Journey.* Michael Wiese Production (3rd edn).
Wilber, Ken (1979). *No Boundary: Eastern and Western Approaches to Personal Growth.* Boston: Shambhala.

Chapter 4

Bullfinch, Thomas (1993). *The Golden Age of Myth and Legend.* Ware: Wordsworth Editions Ltd.
Bunyan, John (1678). *Pilgrim's Progress* (Wordsworth Classics of World Literature, 2009). Wordsworth Editions Ltd.
Conway, D.J. (1996). *Celtic Magic.* St Paul, MN: Llewellyn Publications.
Cooper, J.C. (ed) (2001). *Brewer's Book of Myth and Legend.* Oxford: Helicon Publishing Ltd.
Crowley, Vivienne (1996). *Principles of Paganism.* London: Thorsons.
Dante, Alighieri (1472). *The Divine Comedy* (Wordsworth Classics of World Literature, 2009). Wordsworth Editions Ltd.
Johnson, Revd Nancy (2006). *Our Common Vulnerability.* Lent Sermon, Sheffield.

Rupp, Joyce (2004). *Dear Heart Come Home: the Path of Mid-Life Spirituality.* New York: Crossroad.

Tolkien, J.R.R. (1991). *The Lord of the Rings*. London: Grafton (3 vols, 2nd edn).

Toulson, Shirley (1994). *The Celtic Year.* Shaftesbury: Element.

Von Franz, Marie-Louise (1970 in English). *The Interpretation of Fairy Tales.* London: Shambhala (1996 revised edition).

Chapter 5

Claxton, Guy (2006). *The Wayward Mind: An Intimate History of the Unconscious.* London: Abacus.

Damasio, Antonion (2000). *The Feeling of What Happens: Body, Emotion and the Making of Consciousness.* London: Vintage.

Ferris, Timothy (1992). *The Mind's Sky: Human Intelligence in a Cosmic Context.* London: Bantam.

Frankl, Victor (1946). *Man's Search for Meaning.* (1985 edition) New York: Washington Square Press.

Holmes, Jeremy (1994). *John Bowlby and Attachment Theory.* London: Routledge.

Howe, David (1995). *On Being a Client: Understanding the Process of Counselling and Psychotherapy.* London: Sage.

Le Doux, Joseph (1994). *The Emotional Brain*. London: Phoenix.

Ramachandran, V.S. (2005). *Phantoms in the Brain.* London: Harper Perennial.

Wegner, David (2002). *The Illusion of Conscious Will.* London: MIT Press.

Winnicott, D.W. (1958). *Through Paediatrics to Psychoanalysis.* (1992 edition) London: Karnac Books.

Yalom, Irvin (1991). *Love's Executioner and Other Tales of Psychotherapy.* London: Penguin.

Chapter 6

www.consciousaging.com

www.eomega.org

Atchley, R.C. *Conscious Aging: Nurturing a New Vision of Longevity – But is it a Hard Sell?* American Society on Aging – http://asaging.org/at/at-231/Conscious.html

Austad, Steve (1997). *Why We Age.* New York: John Wiley & Sons Inc.

Austad, Steven (1997). Comparative ageing and life histories in mammals. *Exp Gerontol* **32**(1-2): 23-38.

Bryson, Bill (2003). *A Short History of Nearly Everything.* London: Doubleday.

Darwin, Charles (1859). *The Origin of Species by Means of Natural Selection.* (1974 edition) Harmondsworth: Penguin.

Dass, Ram (2000). *Still Here: Embracing Aging, Changing and Dying.* New York: Riverhead Books.

Gould, Stephen Jay (2002). *The Structure of Evolutionary Theory.* Harvard: Belknap Press.

Kirkwood, T.B. (1977). The evolution of ageing. *Nature* **270**: 301-304.

Kirkwood, T.B. (2001). *The End of Age: Why Everything About Ageing is Changing* (Reith Lecture). London: Profile Books.

Kirkwood, T.B. and Austad, S.N. (2000). Why do we age? *Nature* **408**: 233-238.

MacArthur, R.H. and Wilson, E.O. (1967). *The Theory of Island Biogeography*. Princeton University Press, Princeton.

Medawar, P. (1952). *An Unsolved Problem of Biology.* London: HK Lewis.

Moody, H.R. (2000). *Conscious Aging: A New Level of Growth in Later Life.* www.hrmoody.com

Pallen, Mark (2009). *The Rough Guide to Evolution*. London: Rough Guides Ltd.

Smith, John Maynard (1977). *The Theory of Evolution.* Harmondsworth: Penguin (3rd edn).

Chapter 7

Allal, N., Sear, R., Prentice, A.M., Mace, R. (2006). *The Evolutionary Ecology of Reproduction in Rural Gambia.* The Human Evolutionary Ecology Group, University College London.

Boehm, Christopher (2012). *Moral Origins: the Evolution of Virtue, Altruism and Shame.* New York: Basic Books.

Hawkes, Kristin, O'Connell, J.F., Blurton Jones N.G., Alvarez, H., Charnoir, E.L. (1998). Grandmothering, menopause and the evolution of human life histories. *Proc Natl Acad Sci USA* **95:** 1336-1339.

Lahdenpera, M., Lummaa, V., Helle, S., Tremblay, M., Russell, A.F. (2004). Fitness benefits of post-reproductive lifespan in women. *Nature* **428**: 178-181.

Leacock, Eleanor Burke (1981). *Myths of Male Dominance*. (2008 edition) Chicago: Haymarket Books.

Leakey, Richard and Lewin, Roger (1993). *Origins Reconsidered: In Search of What Makes Us Human.* London: Abacus.

Mithin, Steven (2003). *After the Ice Age: A Global Human History, 20,000-5,000 BC.* London: Phoenix.

Packer, Craig (1998). Why menopause? *Natural History* **107** (6): 24-26.

Peccei, J.S. (1995) A hypothesis for the origin and evolution of menopause. *Maturitas* **21** (2): 83-89.

Roberts, Alice (2009). *The Incredible Human Journey: the Story of How We Colonised the Planet*. London: Bloomsbury.

Shanley D.P. and Kirkwood T.B. (2001). Evolution of the human menopause. *Bioessays* **23**: 282-287.

Voland, E., Chasiotis, A, Schiefenhovel, W. (eds) (2005). *Grandmotherhood: The Evolutionary Significance of the Second Half of a Female Life.* New Brunswick, New Jersey and London: Rutgers University Press.

Chapter 8

Boehm, Christopher (2012). (see above, Chapter 7)

Balter, Michael (2005). *The Goddess and the Bull – Catalhoyuk: an Archaeological Journey to the Dawn of Civilisation.* New York: Free Press.

Bowles, Samuel (2011). History lesson from the first farmers. *New Scientist* 30 July 2011.

Cauvin, Jacques (2000). *The Birth of the Gods and the Origins of Agriculture.* Cambridge University Press (2nd edn).

Engels, Friedrich (1884). *Origin of the Family, Private Property and the State.* (2004 edition) Resistance Books.

German, Lindsey (1989). *Sex, Class and Socialism.* London: Bookmarks.

Gilbert, P., Bailey, K.G. (2000). *Genes On the Couch: Explorations in Evolutionary Psychotherapy.* Hove: Brunner-Routledge.

Hodder, Ian (1990). *The Domestication of Europe.* Blackwell.

Leacock, Eleanor Burke (2008). (See above, chapter 7)

Mann, C.C. (2011). The birth of religion. *National Geographic* June 2011.

Mithin, Steven (1998). *The Prehistory of the Mind.* London: Phoenix.

Mithin, Steven (2003). (See above, Chapter 7)
Stevens, A., Price, J. (2000). *Evolutionary Psychiatry*. London: Routledge.
Renfrew, C. (2007). *Prehistory: Making the Human Mind.* London: Weidenfeld and Nicolson.
Renfrew, C., Zubrow, I. (1994) *The Ancient Mind: Elements of Cognitive Archaeology.* Cambridge University Press.
Roux, Georges (1992). *Ancient Iraq.* London: Penguin (3rd edn).
Taylor, Timothy (1997). *The Prehistory of Sex.* London: Bantam.
Taylor, Timothy (2002). *The Buried Soul: How Humans Invented Death.* Fourth Estate.
Wood, Michael (2005). *In Search of the First Civilisations.* London: BBC Books.

Chapter 9

Baker, R., Bellis, M. (1995). *Human Sperm Competition.* New York: Chapman Hall.
Balter, Michael (2005). (See above, Chapter 8)
Eisler, Riane (1987). *The Chalice and the Blade: Our History, Our Future.* San Francisco: Harper Collins.
Gimbutas, Marija (1982). *The Goddesses and Gods of Old Europe7,000-3500 BC: Myths and Cult Images.* London: Thames and Hudson (2nd edn).
Mellaart, James (1967). *Catalhoyuk, a Neolithic Town in Anatolia.* London: Thames and Hudson.
Stone, Merlin (1976). *The Paradise Papers.* London: Virago.

Chapter 10

Bolen, Jean Shinola (2001). *Goddesses For Older Women: Archetypes in Women Over Fifty.* New York: Harper Collins.
Doherty, Lilian (2001). *Gender and the Interpretation of Classical Myth.* London: Duckworth.
Edwards, Gill (1991). *Stepping Into the Magic.* London: Piatkus.
Hooke, S.H. (1971). *Middle Eastern Mythology.* London: Penguin.
Meador, Betty DeShong (2000). *Inanna, Lady of Largest Heart: Poems of the Sumerian High Priestess Enheduanna.* Austin: University of Texas Press.
Roux, Georges (1992). (See above, Chapter 8)

The Electronic Text Corpus of Sumerian Literature (2006). *The Cursing of Agade – text.2.1.5.* The ETCSL Project, Faculty of Oriental Studies, University of Oxford.
Wood, Michael (2005). (See above, Chapter 8)

Chapter 11

Berggren, Kristin and Harrod, J.B. (1996). Understanding Marija Gimbutas. *Journal of Prehistoric Religions* **X**: 70-73.
Eller, Cynthia (2000). *The Myth of Matriarchal Prehistory – Why an Invented Past Won't Give Women a Future.* Boston: Beacon Press.
Fleming, Andrew (1969). The myth of the mother goddess. *World Archaeology* **1**(2): 247-261.
Goodison, Lucy and Morris, Christine (eds) (1998). *Ancient Goddesses: the Myths and the Evidence.* London: British Museum Press.
Mann, C.C. (2011). (See above, Chapter 8)
Maskell, Lynn (1998). In: Goodison and Morris (1998).
Mellaart, James (1967). (See above, Chapter 9)
Mithen, Steven (2003). (See above, Chapter 7)
Renfrew, Colin (1987). *Archaeology and Language: The Puzzle of Indo-European Origins.* London: Pimlico.
Roux, Georges (1992). (See above, Chapter 8)
Taylor, Timothy (1997). (See above, Chapter 8)

Chapter 12

Armstrong, Karen (1994). *A History of God.* London: Mandarin Paperbacks.
Dever, W.G. (2005). *Did God Have a Wife? Archaeology and Folk Religion in Ancient Israel.* Cambridge: Eerdmans.
Finkelstein, Israel and Silbermann, Neil Asher (2002). *The Bible Unearthed.* New York: Touchstone.
Friedman, Richard Elliott (1988). *Who Wrote the Bible.* London: Jonathan Cape.
Koltuv, B.B. (1986). *The Book of Lilith.* Berwick, Maine: Nicolas-Hays Inc.
Marcus, Amy Docker (2000). *Rewriting the Bible: How Archaeology is Reshaping History.* London: Little, Brown.
Roux, Georges (1992). (See above, Chapter 8)

Spong, John Shelby (1992). *Born of a Woman: a Bishop Rethinks the Virgin Birth and the Treatment of Women by a Male-Dominated Church.* San Francisco: Harper.

Stone, Merlin (1976). (See above, Chapter 9)

Thompson, T.L. (1999). *The Bible in History: How Writers Create a Past.* London: Jonathan Cape.

Warner, Marina (2000). *Alone of All Her Sex: The Myth and Cult of the Virgin Mary.* London: Vintage.

Chapter 13

Christ, Carol P. (1997). *Rebirth of the Goddess.* Abingdon: Routledge.

Graves, Robert (1948). *The White Goddess.* (1999 edition) London: Faber & Faber.

Leeming, D. and Page, J. (1994). *Goddess: Myths of the Female Divine.* Oxford University Press.

Moorey, Teresa (2003). *The Goddess.* London: Hodder and Stoughton.

Walker, Barbara G. (1987). *The Skeptical Feminist: Discovering the Virgin, Mother and Crone.* San Francisco: Harper and Row.

Chapter 14

www.angelfire.com/mi/enheduanna

Binkley, Roberta (1998). *Biography of Enheduanna, Priestess of Inanna* and *Enheduanna: The Overview of her Writings.* Feminist Theory website hosted by the Center for Digital Discourse and Culture at Virginia Tech University (http:// cddc.vt.edu/feminism/enheduanna.html).

Bolen, Jean Shinola (2001). (See above, Chapter 10)

Hallo, W.W. and Van Dijk J.J.A. (trans) (1968). *The Exhaltation of Inanna.* New Haven: Yale University Press.

Leeming D. and Page J. (1994). (See above, Chapter 13)

Meador, Betty DeShong (2000). (See above, Chapter 10)

Moorey, Terea (2003). (See above, Chapter 13)

Perera, Sylvia Brinton (1981). *Descent to the Goddess: A Way of Initiation for Women.* Toronto: Inner City Books.

Sjoberg, A. and Bermann, E. (trans) (1969). *The Collection of the Sumerian Temple Hymns.* Locust Valley, NJ: JJ Augustus.

Wolkenstein, Diane and Kramer, Samuel (1983). *Inanna: Queen of Heaven and Earth.* New York: Harper and Row.

Chapter 15

www.minniepauz.com
www.power-surge.com
Blackmore, S. (1993). *Dying to Live: Science and the Near-Death Experience.* London: Grafton.
Campling, F. and Sharpe, M. (2006). *Living With a Long-Term Illness: The Facts.* Oxford University Press.
Charmaz, Kathy (1983). Loss of self: a fundamental form of suffering in the chronically ill. *Sociology of Health and Illness* **5** (2).
Greer, Germaine (1991). *The Change: Women, Ageing and the Menopause.* London: Hamish Hamilton.
Harrison, Ted (2001). *Beyond Dying.* Oxford: Lion Books.
Holroyd, S. (1991). *Krishnamurti: the Man, the Mystery and the Message.* Shaftesbury: Element.
Le Maistre, Joan (1995). *After the Diagnosis: From Crisis to Personal Renewal For Patients With Chronic Illness.* Berkeley, CA: Ulysses Press.
Moore, Thomas (2004). *Dark Nights of the Soul.* London: Piatkus.
Murray, Jenni (2003). *Is It Me Or Is It Hot In Here?* London: Vermilion.
Nuland, Sherwin B. (1994). *How We Die.* London: Chatto and Windus.
Senior, M. with Viveash, B. (1998). *Health and Illness.* London: MacMillan.
Taylor, Timothy (2002). (See above, Chapter 8)
Ramachandran, V.S. (2005). (See above, Chapter 5)
Worth, Jennifer (2011). *In the Midst of Life.* London: Phoenix.

Chapter 16

Borg, Marcus (2004). *The Heart of Christianity.* San Francisco: Harper.
Ferris, Timothy (1992). (See above, Chapter 5)
Geering, Lloyd (2000). *Tomorrow's God: How We Create Our Worlds.* Polebridge Press.
Gilbert, R.A. (1991). *The Elements of Mysticism.* Shaftesbury: Element Books.
Huxley, Aldous (1946). *The Perennial Philosophy.* Chatto and Windus.
Kopp. Sheldon (1993). *If You Meet the Buddha On the Road, Kill Him.* London: Sheldon Press.

Jones, James W. (1995). *In the Middle of This Road We Call Our Life.* London: Thorsons.

Merton, Thomas (1993). *No Man Is An Island.* Tunbridge Wells: Burns and Oates.

Pascal, Blaise (2001). *Religion Explained: the Human Instincts that Fashion Gods, Spirits and Ancestors.* London: Wm Heinemann.

Peck. M. Scott (1993). *Further Along the Road Less Travelled.* New York: Simon and Schuster.

Rinpoche, Sogyal (1992. *The Tibetan Book of Living and Dying.* London: Rider.

Sagan, Carl with Druyan, Ann (1997). *Billions and Billions: Thoughts on Life and Death at the Brink of the Millenium.* Ballantine Books.

Winston, Robert (2005). *The Story of God.* London: Bantam Press.

Yalom, Irvin (2008). *Staring Into the Sun: Overcoming the Dread of Death.* London: Piatkus.

Chapter 17

Basham, A.L. (1954). *The Wonder That Was India.* New York: Grove Press.

Meador, Betty DeShong (2000). (See above, Chapter 10)

Roux, Georges (1992). (See above, Chapter 8)

Chapter 18

Berne, Eric (1970). *What Do You Say After You Say Hello?* (1994 edition) London: Corgi.

Edwards, Gill (1991). (See above, Chapter 10)

Ernst, Sheila and Goodison, Lucy (1981). *In Our Own Hands: A Book of Self-Help Therapy.* London: The Women's Press.

George, E., Iveson, C., Ratner, H. (1999). *Problem to Solution: Brief Therapy With Individuals and Families.* London: Brief Therapy Press (2nd edn).

Gibran, Kahlil (1923). *The Prophet.* (Wordsworth Classics of World Literature, 1997). Wordsworth Editions Ltd.

Griffin, J. and Tyrell, I. (2006). *How to Lift Depression – Fast.* Chalvington: Human Givens Publishing Ltd.

Mearns, D. (1995). *Developing Person-Centred Counselling.* London: Sage.

Medici, Marina (1988). *Good Magic.* London: MacMillan.

Oliver, James (2002). *They F**k You Up: How to Survive Family Life.* London: Bloomsbury.
Schiffman, M (1990). *Gestalt Self Therapy.* Berkeley: Wingbow Press.
Shone, Neville (2002). *Coping Successfully With Pain.* London: Sheldon Press.
Siegel, Bernie (1990). *Love and Healing.* New York: Harper Perennial.
Simonton, O.C., Matthews-Simonton, S., Creighton, J.L. (1992). *Getting Well Again.* New York: Bantam.
Summers, C. and Vayne, J. (2002). *Personal Development With the Tarot.* London: Quantum.

Chapter 19

Bates, Brian (1996). *The Wisdom of the Wyrd.* London: Ryder.
Bolen, Jean Shinola (2001). (See above, Chapter 10)
Conway, D.J. (1996). (See above, Chapter 4)
Hart, Eloise (2006). *Kuan Yin: Goddess of Mercy, Friend of Mankind.* www.theosophy-nw.org/theosnw/world/asia/as-elo.htm.
Leeming D. and Page J. (1994). (See Chapter 13)
Wikipedia.

Correspondence

To enquire about workshops, and for all other correspondence, please email me at: contact@stone103.plus.com.

www.ingramcontent.com/pod-product-compliance
Ingram Content Group UK Ltd.
Pitfield, Milton Keynes, MK11 3LW, UK
UKHW041946190726
13854UKWH00004B/1823